Environmental Science and Engineering

The ultimate goal of this series is to contribute to the protection of our environment, which calls for both profound research and the ongoing development of solutions and measurements by experts in the field. Accordingly, the series promotes not only a deeper understanding of environmental processes and the evaluation of management strategies, but also design and technology aimed at improving environmental quality. Books focusing on the former are published in the subseries Environmental Science, those focusing on the latter in the subseries Environmental Engineering.

Chaolei Yuan · Suiliang Huang · Xin Wang ·
Zeyou Chen
Editors

Proceedings of 4th International Conference on Resources and Environmental Research—ICRER 2022

Springer

Editors
Chaolei Yuan
School of Agriculture
Sun Yat-sen University
Shenzhen, China

Xin Wang
College of Environmental Science
and Engineering
Nankai University
Tianjin, China

Suiliang Huang
College of Environmental Science
and Engineering
Nankai University
Tianjin, China

Zeyou Chen
College of Environmental Science
and Engineering
Nankai University
Tianjin, China

ISSN 1863-5520 ISSN 1863-5539 (electronic)
Environmental Science and Engineering
ISBN 978-3-031-31807-8 ISBN 978-3-031-31808-5 (eBook)
https://doi.org/10.1007/978-3-031-31808-5

This Springer imprint is published by the registered company Springer Nature Switzerland AG
The registered company address is: Gewerbestrasse 11, 6330 Cham, Switzerland

Paper in this product is recyclable.

Committees

Conference Honorary Chair

Hongwen Sun, Nankai University, Tianjin, China

Conference General Chair

Suiliang Huang, College of Environmental Science and Engineering, Nankai University, Tianjin, China

Conference Co-Chairs

Xin Wang, College of Environmental Science and Engineering, Nankai University, Tianjin, China
Gordon Huang, University of Regina, Canada

Conference Program Chair

Yongping Li, Beijing Normal University, China

Conference Publication Chairs

Chaolei Yuan, Sun Yat-sen University, Shenzhen, China
Zhen Hu, Shandong University, China

Guest Editors

Xin Wang, College of Environmental Science and Engineering, Nankai University, Tianjin, China
Zeyou Chen, College of Environmental Science and Engineering, Nankai University, Tianjin, China

Conference Steering Committees

Mohamed El-Din, University of Alberta, Canada
Satya Harpalani, Southern Illinois University Carbondale, USA

Student Competition Chairs

Mark Gerrard Bartlam, Nankai University, Tianjin, China
Pier-Luc Tremblay, Wuhan University of Technology, China
DONG Yahong Caroline, Macau University of Science and Technology, China
Oh Wen Da, Universiti Sains Malaysia, Malaysia

Technical Committee

Zhen Hu, Shandong University, China
Eric James Strauss, Michigan State University, USA
Hasan Aydogan, Istanbul University, Turkey
Ali Cheshmehzangi, University of Nottingham Ningbo, China
Saad Abdel-Hamid El-Syaed, Mechanical power engineering-Zagazig University, Egypt
Yanfeng Li, South-South Cooperation Center for Sustainable Development, China
Rodrigo C. Munoz, Bataan Peninsula State University, Philippines
Bal Deshwal, A. I. Jat H. M. College, India
Bandar Alqahtani, Duke University, USA

Preface

The 2022 4th International Conference on Resources and Environmental Research (ICRER 2022) was successfully held online (due to the COVID-19 pandemic) from 2 to 4 December 2022. ICRER 2022 was organized by Nankai University, and technically supported by the Xiamen University of Technology, University of Nottingham Ningbo, and Qatar University.

The Conference was attended by around 80 researchers and students from China, USA, Mexico, Japan, UK, Australia, Canada, Singapore, Sweden, Belgium, Germany, Poland, Turkey, Vietnam, Thailand, the Philippines, etc. The Conference included Normal Oral Sessions and Student Competition Sessions. Forty-five graduate students competed for ten Best Young Oral Presenters sponsored by the Royal Society of Chemistry, UK.

The Conference was highlighted by 18 keynote speakers: (1) Prof. Zhaoyin Wang from Tsinghua University, China; (2) Prof. Gustavo Santoyo from the National Autonomous University of Mexico, Mexico; (3) Prof. (Neal) Tai-Shung Chung from the National University of Singapore, Singapore; (4) Prof. Xu Zhihong from Griffith University, Australia; (5) Prof. Miklas Scholz from Lund University, Sweden; (6) Prof. Korneel Rabaey from Ghent University, Belgium; (7) Prof. Rainer Helmig from the University of Stuttgart, Germany; (8) Prof. Emma Eley from Royal Society of Chemistry, UK; (9) Prof. Shaobin Wang from the University of Adelaide, Australia; (10) Prof. Yu-You Li from Tohoku University, Japan; (11) Prof. Gökmen Tayfur from the İzmir Institute of Technology, Turkey; (12) Prof. Adam Szymkiewicz from the Gdańsk University of Technology, Poland; (13) Prof. Xinmin Zhan from the National University of Ireland, Galway, Ireland; (14) Prof. Kurunthachalam Kannan from New York University School of Medicine, USA; (15) Prof. Hui Li from Michigan State University, USA; (16) Assoc. Prof. Luke Mosley from the University of Adelaide, Australia; (17) Assoc. Prof. Romanovski Valentin from the National University of Science and Technology, Belarus; and 18) Dr. Bertram Boehrer from the Helmholtz Centre for Environmental Research—UFZ, Germany.

The Conference also had three invited speakers: (1) Prof. Gang Han from Nankai University, China; (2) Assoc. Prof. Chaolei Yuan from Sun Yat-sen University, China;

and (3) Dr. Qian Zhang from the University of Maryland Center for Environmental Science, USA.

The emphasis of the Conference was on the effective use of environmental resources, ecological restoration and remediation of degraded and polluted environment, and emission reduction technologies of environmental energy system. All papers submitted to the Conference were peer-reviewed. Accepted papers are included in this volume, which comprises four chapters: (1) Analysis and Treatment of Water Pollution, (2) Meteorology and Bioclimatic Design, (3) Ecological Environment and Resource Management, and (4) Traffic Management and Related Carbon Emission Analysis.

We offer our sincere gratitude to the Organizing Committee of the Conference, the Technical Committee of the Conference, and the volunteers, who had dedicated their time and efforts to planning, organizing, and promoting the conference. Special thanks are given to all session chairs for their great support.

Shenzhen, China
January 2023

Chaolei Yuan

Contents

Part I
Analysis and Treatment of Water Pollution

Chapter 1
Study on the Correlation Between the Variation of DO and pH Value and Chlorophyll-a Concentration in Water Body

Z. Lin, M. L. Zhu, J. C. Tu, H. Jiang, and Y. L. Han

Abstract Physiological activities of algae have periodic effects on DO and pH in water. In order to improve the early warning ability of algal blooms, it is necessary to carry out study on the correlation between DO, pH and Chl-a concentration. In this study, a 48-h day and night observation of DO, pH and water temperature were carried out for three study water bodies. At the same time, the concentration of Chl-a was monitored once a day, and the variation rules of DO, pH and water temperature in three study water bodies, as well as the correlation between the diurnal variation of DO and pH and the concentration of Chl-a were analyzed. The results showed that the DO, pH and water temperature of the three water bodies showed a significant diurnal variation pattern. The lowest and highest values of DO, pH and water temperature in the upper water appeared at 6:00 and 14:00–16:00, respectively; The maximum value of the lower water lags behind that of the upper water, from 18:00 to 20:00, and the higher the concentration of Chl-a, the more significant the lag is; In addition, the diurnal variation amplitudes of DO and pH (ΔDO, ΔpH) have a significant linear positive correlation with the concentration of Chl-a. The linear fitting relationship between Chl-a(y) concentration and ΔDO(x) is $y = 8.697x + 9.160$, and the coefficient of determination is 0.879; the relationship between Chl-a(y) concentration and ΔpH(x) is $y = 86.885x - 56.766$, the coefficient of determination is as high as 0.965. The results can provide a simple and effective method for early warning of algal blooms by monitoring the diurnal changes of pH in water bodies to estimate algae concentrations for field work that lacks monitoring conditions for Chl-a concentration.

Keywords DO · pH · Chl-a · ΔDO · ΔpH · Correlation

Z. Lin · M. L. Zhu (✉) · J. C. Tu · H. Jiang · Y. L. Han
School of Environmental Science and Engineering, Xiamen University of Technology, Xiamen 361024, China
e-mail: 292890310@qq.com

The Key Laboratory of Water Resources Utilization and Protection of Xiamen, Xiamen University of Technology, Xiamen 361024, China

C. Yuan et al. (eds.), *Proceedings of 4th International Conference on Resources and Environmental Research—ICRER 2022*, Environmental Science and Engineering,
https://doi.org/10.1007/978-3-031-31808-5_1

1.1 Introduction

The acceleration of urbanization and the massive discharge of domestic sewage have led to aggravated eutrophication of various small and micro water bodies, rivers and lakes, resulting in frequent outbreaks of cyanobacterial blooms. In recent years, all parts of the country have been actively promoting the prevention and control of algal blooms. In December 2021, the National Development and Reform Commission issued the "14th Five-Year Plan" Comprehensive Water Environment Management Plan for Key River Basins, which clearly calls for strengthening the eutrophication of lake and reservoir water sources, improve the early warning ability of cyanobacteria blooms.

Nitrogen and phosphorus nutrients are considered to be the main factors causing cyanobacterial blooms, but the response relationship between the two and cyanobacterial blooms is quite different due to the influence of different environmental factors (Lee et al. 2015). At present, some scholars at home and abroad believe that when the concentration of nitrogen and phosphorus in water exceeds 0.5 and 0.02 mg/L, and the concentration of chlorophyll-a exceeds 10 μg/L, it is easy to cause cyanobacterial blooms (Cao et al. 2022; Kong and Gao 2005). The formation of cyanobacterial blooms is the result of the dominance of cyanobacteria in the succession of phytoplankton communities, changes in many environmental factors have previously predicted this outcome. Previous studies have shown that the physiological activities of algae will affect the dissolved oxygen (DO) and pH in normal or eutrophic water (Ji et al. 2013). Huang et al. (2011) found a significant correlation between the changes of DO and pH and the density of algae in the long-term observation of natural enclosed water bodies and lakes. The diurnal observations of Qi et al. (2018) in Qiusuo Creek, Yichang showed that phytoplankton and water temperature were the main factors affecting the diurnal variation of DO and pH. In addition, the concentration of chlorophyll-a is significantly positively correlated with the number of algae, and the concentration of chlorophyll-a (Chl-a) in a water can roughly reflect the situation of algae in water (Xun et al. 2011). At present, many scholars have analyzed the feasibility of predicting algal blooms based on Chl-a concentration (Wu et al. 2021; Shi et al. 2022; Zhu et al. 2020). It can be seen that by establishing the quantitative relationship between DO, pH and Chl-a, it is expected to evaluate the growth status of algae based on the changes of DO and pH in the water body, and then evaluate the degree of eutrophication in the water body and early warning of algal blooms. At present, scholars have carried out related study on the diurnal variation ΔDO of DO. For example, You et al. (2007) predicted the growth of algae by the relationship between ΔDO and chlorophyll-a concentration. Wang et al. (2018) used this relationship to make a Criteria for evaluating water eutrophication. However, the existing studies are mainly based on artificial water bodies configured in the laboratory, or long-term observations of an actual water body to analyze and explore the relationship between Chl-a concentration, DO and pH. Comparative study on water bodies is lacking in discussing the differences between the upper and lower layers of water bodies, and it is even rarer to conduct study on the quantitative relationship between the diurnal variation of pH (ΔpH) and the concentration of Chl-a in water.

For this reason, this paper takes the Xiamen University of Technology Polytechnic Lake and the two ponds in Houtang Village, Tongan District, Xiamen as study objects. The DO, pH and water temperature of these three water bodies were monitored for a total of 48 h for two days, day and night monitoring every 2 h, and Chl-a concentration was monitored once a day. Based on the monitoring data, the changes of DO, pH and water temperature in the three study water bodies were analyzed, and established the correlation formula of ΔDO, ΔpH and Chl-a concentration. Therefore, it provides a convenient and effective method for the future field work to simply evaluate the algal concentration and the risk of algal blooms in the water body based on the DO and pH observations.

1.2 Study Objects and Methods

1.2.1 Study Object

The study objects are three water bodies located in Xiamen City. Xiamen has a subtropical marine monsoon climate, with high temperatures, high temperature days and persistent high temperatures. The annual average temperature is 21.9 °C, and the annual precipitation is 1123.0 mm. The overview of the water bodies of the three study objects is as follows:

The 1# water body is Xiamen University of Technology Polytechnic Lake (N24°37′36″, E118°5′14″), located in Jimei Campus of Xiamen University of Technology, covering an area of about 6200m^2, with an average water depth of 1.7 m. The water source mainly depends on rainwater runoff and groundwater recharge. The 2# water body is a pond beside the fish pond (N24°45′55″, E118°11′7″) in Houtang Village, Tongan District, Xiamen City. It covers an area of 820 m^2 and has an average water depth of 1.5 m. The pond's water supply relies on rainwater and domestic sewage discharged by a small number of residents. The 3# water body is also located in Houtang Village, Tongan District, called the Shuiwei Palace Front Pond (N24°45′84″, E118°11′15″), covering an area of 340 m^2 and an average water depth of 1.3 m. Water supply also relies on rainwater and a small amount of domestic sewage.

1.2.2 Methods

In this study, the online automatic monitoring of DO, pH and water temperature in the upper and lower water layers for 2 days and 48 h was carried out in three study water bodies. The monitoring water depths of the upper and lower water layers were

Table 1.1 Instruments used to monitor water quality indicators

Water quality index	Instrument	Supplier	Method
pH, DO, Water temperature	Water quality index online automatic monitor	Qingdao Bangbang Technology Co., Ltd.	Membrane electrode and glass electrode
Chlorophyll-a	PHYTO-PAM	Germany WALZ company	Fluorescence

set at 10 cm (upper water layer) and 65 cm (lower water layer) respectively, and the automatic monitoring time interval is 2 h. In addition, the concentration of Chl-a was detected by manual sampling from 14:00 to 15:00 every day. The monitoring of 1# water body (Polytechnic Lake) was from 14:00 on July 10, 2022 to 14:00 on July 12, 2022, which lasted for a total of 48 h; The monitoring period of 2# water body (pond beside the fish pond) was from 14:00 on July 28, 2022 to 14:00 on July 30, 2022, which lasted for a total of 48 h; The monitoring of the 3# water body (Shuiwei Palace Front Pond) was from 14:00 on June 30, 2022 to 14:00 on July 2, 2022, for a total of 48 h. The instruments and detection methods used for monitoring various water quality indicators are shown in Table 1.1.

1.3 Results and Discussion

1.3.1 Comparison of Algae Content in Three Study Water Bodies

In this study, Chl-a was monitored continuously for 24 h for 1# (Polytechnic Lake) (3 water samples were taken every 2 h to detect the concentration of Chl-a), and it was found that the concentration of Chl-a remained basically stable during the day and night (between 22 to 23.5 μg/L). Based on the diurnal stability of Chl-a and the fact that the 2# and 3# water bodies are far away from the school laboratory, it is difficult to carry out continuous detection of Chl-a concentration. Therefore, this study is aimed at three water bodies, and the pH, DO and water temperature are continuously measured for 48 h. During the monitoring period, multiple water samples were collected for Chl-a detection only from 14:00 to 15:00 every day, instead of continuous monitoring. The test results are shown in Table 1.2.

It can be seen from Table 1.2 that the algae content of the 1# water body is the least, and its average Chl-a concentration is only 22.57 μg/L, followed by the 2# water body, which average Chl-a concentration is 103.43 μg/L, and the 3# water body has the highest concentration, which is 129.45 μg/L.

Table 1.2 Chl-a concentration during monitoring of three water bodies

Study object	Sampling time	Chl-a (μg/L)	Average (μg/L)
1# (Polytechnic lake)	2022/7/10 14:25	22.69	22.57
	2022/7/11 14:15	22.44	
2# (Pond beside the fish pond)	2022/7/28 14:10	90.51	103.43
	2022/7/29 14:35	116.35	
3# (Shuiwei palace front pond)	2022/6/30 14:05	126.05	129.45
	2022/6/30 14:38	132.84	

1.3.2 Analysis of Diurnal Variation of DO, pH and Water Temperature

Because it was in summer and affected by subtropical high, the average temperature of the three study water bodies during the observation period was about 32 °C during the day and 25 °C at night. The observation results of DO, pH and water temperature of the three study water bodies 1#, 2#, and 3# are shown in Fig. 1.1a, b and c, respectively. As can be seen from Fig. 1.1.

The DO, pH and water temperature in the upper and lower layers of the three studied water bodies all have obvious 24-h changes. The DO, pH and water temperature of the upper layer are at the minimum value in one day at 6:00 during the day, and then continue to rise, with a high rate of rise between 8:00 and 14:00, and at 14:00–18:00 in the afternoon reach the highest value of the day and then start to gradually decrease. It can be seen that there is a significant difference in the rate of DO decline in the upper layer of 1# water body and 2# and 3# water bodies after reaching the highest value. The DO of 2# and 3# water bodies decreases rapidly, while the DO of 1# water body fluctuate around the highest value for 6 to 8 h and then drops abruptly between 22:00 and 0:00. The reason why the DO doesn't drop rapidly in the upper layer of 1# water body may be related to the light intensity and the concentration of algae. The strong sunlight in the afternoon inhibits the photosynthesis of the algae in the upper layer of the 1# water body, which reduce the O_2 produced by the algae. After 16:00, with the decrease of the light intensity, the photosynthesis gradually recovered. This can also be confirmed from the fact that the DO in the lower layer of the 1# water body is higher than that in the upper layer. In addition, the content of algae in the 1# water body is less, and the amount of O_2 consumed by respiration is equivalent to the amount of O_2 produced by photosynthesis, resulting in a stable fluctuation of DO within 6–8 h, and the overall downward trend of DO is delayed.

The changes of DO, pH and water temperature in the lower layer of the three study water bodies are basically the same as those in the upper layer. The minimum value in one day appears at 6:00, but the time of the highest value appears lags behind that of the upper layer, about 18:00–20:00, and the highest value of the lower layer is often smaller than the highest value of the upper layer. The difference between the upper and lower layers is mainly related to the phototaxis of algae. The study

Fig. 1.1 The diurnal variation of DO, pH and water temperature during the monitoring period of the three study water bodies

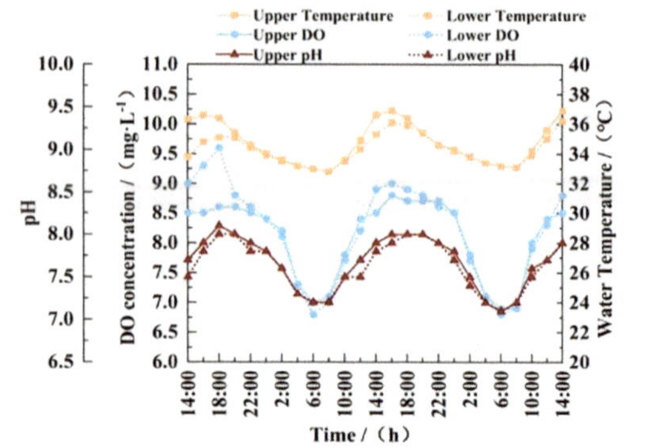

(a) 1# DO, pH and water temperature diurnal variation diagram

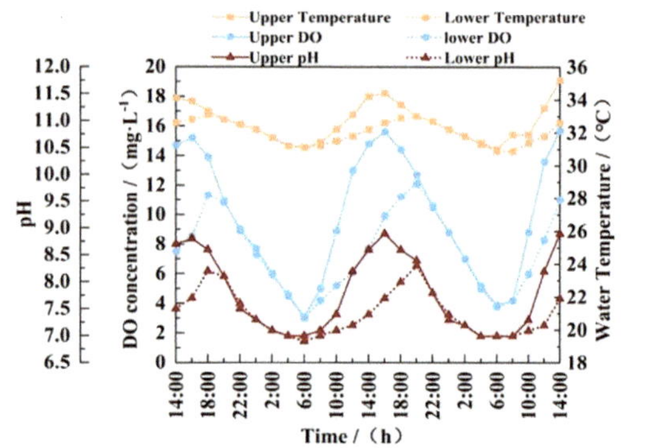

(b) 2# DO, pH and water temperature diurnal variation diagram

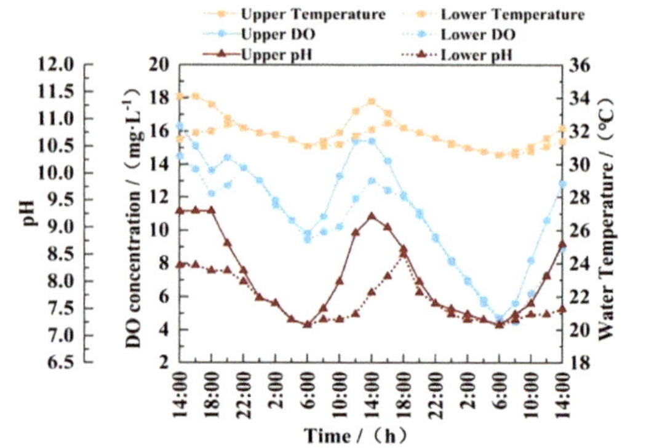

(c) 3# DO, pH and water temperature diurnal variation diagram results of Liu et al. (2015) showed that algae would appear to float significantly under the induction of light, and with the increase of light intensity, the higher the concentration of algae in the upper layer of the water body. Therefore, when the sunlight gradually increases during the day, the algae in the water body moves to the upper layer, resulting in higher DO and pH in the upper layer than that in the lower layer. After the light intensity decreases at night, the algae no longer gather in the upper layer and spread to the deep layer, and at this time, the DO and pH in the upper and lower layer gradually tend to be consistent. In addition, it can be clearly seen from Fig. 1.1 that the difference between the upper and lower layers of DO, pH and water temperature is not obvious in 1# water body, but it is very significant in 2# and 3# water bodies. This is related to the number of aquatic organisms such as algae in the water body. It can be seen from Table 1.2 that the concentration of Chl-a in 1# water body (Polytechnic Lake) is only 22.57 μg/L, while the concentrations of Chl-a in 2# water body (pond beside the fish pond) and 3# water body (Shuiwei Palace Front Pond) are as high as 103.43 and 129.45 μg/L. Therefore, when the concentration of Chl-a in water is smaller, the changes of upper and lower layers are more synchronous, and the peak difference is smaller. On the contrary, the peak time of lower water layer lags behind that of upper water layer, and the peak difference is larger.

The daily fluctuation range of DO in the three study 2# and 3# water bodies is basically the same, but the 1# water body is obviously smaller. During the observation period, the highest value of DO in the upper layer of 1# water body was 8.6 mg/L, the lowest value was 7.0 mg/L, and the fluctuation range was only 1.6 mg/L; the highest value of DO in the upper layer of 2# water body was 15.7 mg/L, and the minimum value was 3.8 mg/L, the fluctuation range was as high as 11.9 mg/L; the highest value of DO in the upper layer of the 3# water body was 16.3 mg/L, the minimum value was 4.7 mg/L, and the fluctuation range also reached 11.6 mg/L. The difference in the fluctuation range may be due to the fact that during the day, with the enhancement of algal photosynthesis, the rate of O_2 production is greater than the consumption rate, resulting in a large increase in DO. After the weakening of photosynthesis at night, the respiration of algae and other organisms consumes a large amount of oxygen, which makes the concentration of DO drop rapidly, thus causing a large fluctuation between day and night.

The change of pH is generally consistent with DO, and the time of peak appearance is synchronized with DO, but there is no rapid rise and fall. During the observation period, the highest value of the upper pH of 1# water body was 8.1, the lowest value was 7.1, and the fluctuation range was 1.0; The maximum value of pH in the upper layer of 2# water body was 9.0, the minimum value was 7.2, and the fluctuation range was 1.8; The highest value of pH in the upper layer of 3# water body was 9.2, the lowest value was 7.2, and the fluctuation range was 2.0. Although the diurnal fluctuation range of pH is small, it is obvious that the fluctuation range of 1# water body is smaller than that of 2# and 3# water bodies. The increase and decrease of pH are mainly related to the carbonate equilibrium system in water, and the contents of CO_2, CO_3^{2-} and HCO_3^- in water are the key factors affecting this equilibrium (Xu et al. 2009). When algae photosynthesis consumes a large amount of CO_2 dissolved in water during the day, it destroys the original balance of H_2CO_3 and causes the pH to rise (López-Archilla et al. 2004). CO_2 released by algae respiration at night dissolves in water to form H_2CO_3, and the ionized CO_3^{2-} lowers the pH of water (Su et al. 2016). Therefore, the number of algae is the main reason for the diurnal fluctuation range of pH in 1# water body less than 2# and 3#.

1.3.3 The Relationship Between ΔDO, ΔpH and Chl-a

From the qualitative analysis in 3.2, it can be seen that the diurnal variation of DO and pH is related to algae activities, and the algae content in 3# water body is the largest, and the diurnal variation range of DO and pH, that is, the larger the ΔDO and ΔpH, the smallest the algae content in 1# water body, and the smallest the ΔDO and ΔpH. To further clarify the correlation between ΔDO and ΔpH and algae content, the following quantitative analysis is made on the correlation between ΔDO and Chl-a and ΔpH and Chl-a. Three water bodies were continuously monitored for two days, and there were $3 * 2 = 6$ data corresponding to the daily Chl-a concentration value (Table 1.2) and the change amplitude ΔDO of the upper layer from peak to trough on that day. Similarly, there are 6 data corresponding to Chl-a concentration and ΔpH. According to these data, the correlation diagrams of ΔDO and Chl-a concentration, ΔpH and Chl-a concentration can be obtained as shown in Figs. 1.2 and 1.3.

As can be seen from Fig. 1.2, in the correlation diagram between ΔDO and Chl-a concentration, except for one point (marked as ×), the other five points show good correlation, and the linear fitting result is $y = 8.697x + 9.160$, with a determination coefficient of 0.879. The reason for the abnormality of the × point data may be related to the cumulative 1.2 mm rainfall of the 3# water body at 16:50 on June 30. The short-term rainfall changed the water environment and increased the DO in the water. This is shown in Fig. 1.1 (The rise of DO after the first 18:00 in c) can also be confirmed. In addition, as can be seen from Fig. 1.3, 6 points of ΔpH and Chl-a concentration show very good correlation. The linear fitting result is $y = 86.885x -$

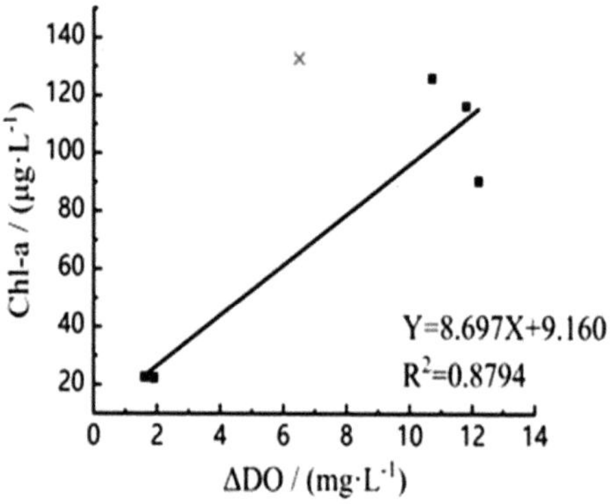

Fig. 1.2 ΔDO and Chl-a concentration correlation

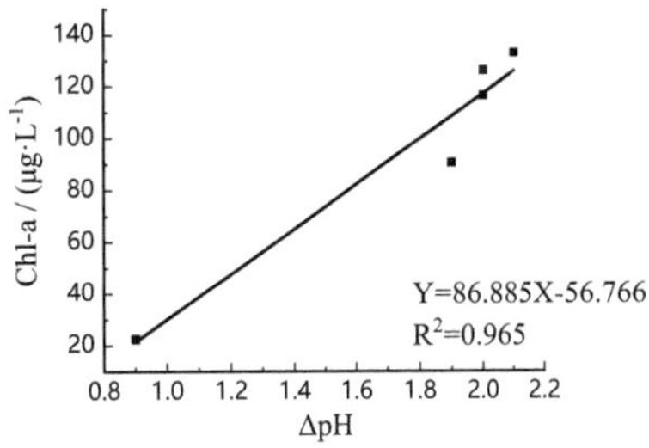

Fig. 1.3 ΔpH and Chl-a concentration correlation

56.766, and the determination coefficient is 0.965. Comparing Fig. 1.2 with Fig. 1.3, it can be seen that the linear correlation between ΔpH and Chl-a concentration is better than ΔDO and Chl-a, and the correlation coefficient is higher.

To sum up, by monitoring the pH value of the upper layer water body at 14:00–16:00 in the afternoon and at 6:00 in the morning, the diurnal variation range of pH in the water body, ΔpH, was calculated, and then the linear simulation between ΔpH and Chl-a concentration was used. Combined with the mathematical formula, the concentration of Chl-a in the water body can be estimated. The study results can provide a simple and effective method for early warning of algal blooms by monitoring the diurnal changes of pH in water bodies to estimate algae concentrations for field work that lacks monitoring conditions for Chl-a concentration.

1.4 Conclusion

In this study, the DO, pH and water temperature of 3 water bodies, 1# (Science and Technology Lake), 2# (Pond near fishpond) and 3# (Pond in front of Shuiwei Palace), were observed day and night for 2 days, and the Chl-a concentration was monitored once a day. By analyzing the variation rules of DO, pH and water temperature in three water bodies, and the correlation between the diurnal variation range of DO, pH and Chl-a concentration, the following conclusions are obtained:

(1) There is an obvious diurnal variation law of DO, pH and water temperature with a 24-h cycle in the three studied water bodies. The DO, pH and water temperature of the upper water body are at the minimum value of one day at 6:00 in the morning, then begin to rise continuously, reach the maximum value in one day between 14:00 and 18:00 in the afternoon, and then begin to gradually decrease. However, the peak time of DO, pH and water temperature in the lower water body lags behind that in the upper water body, and it appears at about 18:00–20:00, and the highest value in the lower water body is smaller than that in the upper layer, but the lowest value appears at 6:00 a.m. like that in the upper layer water body.

(2) The diurnal variation of DO, pH and water temperature are different between upper and lower layers. This difference is not obvious in 1# water body, but it is very obvious in 2# and 3# water bodies, which is related to the number of aquatic organisms such as algae in water bodies. The Chl-a concentration of 1# water body (Polytechnic Lake) was only 22.57 μg/L, while the Chl-a concentration of 2# water body (pond beside the fish pond) and 3# water body (Shuiwei Palace Front Pond) was as high as 103.43 and 129.45 μg/L. The smaller the concentration of Chl-a is, the more synchronous the changes of the upper and lower water layers are, and the smaller the peak difference is. On the contrary, the peak time of the lower water layer lags behind that of the upper water layer, and the larger the peak difference is.

(3) The daily fluctuations of DO and pH, that is, ΔDO, ΔpH, have very good linear positive correlation with the concentration of Chl-a. The linear correlation between ΔpH and Chl-a concentration is more significant than that of ΔDO and Chl-a concentration, and is less susceptible to environmental influences. Therefore, when monitoring the water environment in the field, the Chl-a concentration and the risk of algal blooms can be estimated simply by monitoring ΔpH.

Acknowledgements This work was financially supported by "Fujian Provincial Science and Technology Project (2013I01010022)".

References

Cao J, Tian Z, Chu Z et al (2022) Nitrogen and phosphorus control thresholds of cyanobacterial blooms in lake Taihu. J Lake Sci 34(04):1075–1089

Huang S, Zang C, Du S et al (2011) Study on the relationships among pH, dissolved oxygen and chlorophyll-a II: Non-aquaculture water. Chin J Environ Eng 5(08):1681–1688

Ji X, Li P, Shang X et al (2013) Diurnal variation and evaluation of water quality of the Wen-Rui Tang River in spring. Environ Chem 32(11):2082–2089

Kong F, Gao G (2005) Hypothesis on cyanobacteria bloom-forming mechanism in large shallow eutrophic lakes. Acta Ecol Sin 3:589–595

Lee TA, Rollwagen-Bollens G et al (2015) The influence of water quality variables on cyanobacterial blooms and phytoplankton community composition in a shallow temperate lake. Environ Monit Assess 187(6):315

Liu J, Yang Z, Zhang X et al (2015) Research on vertical distribution characteristics of Melosira with influence of different environmental factors. Mar Sci 39(07):22–28

López-Archilla AI et al (2004) Phytoplankton diversity and cyanobacterial dominance in a hypereutrophic shallow lake with biologically produced alkaline pH. Extremophiles **8**(2):109–115

Qi T, Qin S, Ning Z et al (2018) Diurnal variation characteristics of pH and dissolved oxygen in Qiusuo stream in Yichang in autumn. Yangtze River 49(S2):38–42

Shi H, Chen Q, Zhang Y et al (2022) Temporal and spatial variation of DO and pH and their correlation with chlorophyll A in the northern Bay of Taihu lake. Sichuan Environ 41(02):110–117

Su F, Gao P, Lai Q et al (2016) Effects of Microcystis aeruginosa and Chlorella pyrenoidosa on water environment pH. J Fishery Sci China 23(06):1380–1388

Wang Y, Yang L, Li Z et al (2018) Evaluating eutrophication of landscape water by single index of ADO. China Water Wastewater 34(23):64–68

Wu D, Jia G, Wu H (2021) Chlorophyll-a concentration variation characteristics of the algae-dominant and macro-phyte-dominant areas in Lake Taihu and its driving factors 2007–2019. J Lake Sci 33(05):1364–1375

Xu H, Liu Z, Yuan L et al (2009) Effect of pH on growth of several freshwater algae. Environ Sci Technol 32(01):27–30

Xun S, Yang Y, He B et al (2011) Analysis of relationship between spring water temperature and chlorophyll-a in lake Chaohu. J Lake Sci 23(05):767–772

You L, Cui L, Liu Z et al (2007) Correlation analysis of parameters in algal growth. Environ Sci Technol 09:42–44

Zhu G, Shi K, Li W et al (2020) Seasonal forecast method of cyanobacterial bloom intensity in eutrophic Lake Taihu, China. J Lake Sci 32(05):1421–1431

Chapter 2
Study on Treatment of Compressed Wastewater from Domestic Waste Transfer Station

Liyong Wang, Panfeng Gao, Haiyan Fu, Yuan Dai, and Shuaihao Liu

Abstract Garbage pressure filtrates are pressure filtrates produced when waste is compressed at the transfer station. We mainly analyze the pressure filtrates gencrated during the waste compression process. Combined with the actual situation of waste transfer stations in Xiamen, due to the heavy rainfall in Xiamen, the output of pressure filtrates generally accounts for 10 to 20% of the waste transfer volume, the pressure filtrate cannot be directly discharged into the municipal pipe network due to the high degree of pollution, it shall be treated and prevent secondary pollution caused by the pressure filtrate. In order to meet the Xiamen city's three-level discharge standard for water pollution, the technology and plan for the treatment of waste pressure filtrate are discussed. This process mainly uses PAC-PAM flocculation and precipitation, mineral powder to remove total phosphorus, ammonia nitrogen remover to remove ammonia nitrogen and other methods. Results show that pH 7.00, PAM 2 mg/L, PFC 2 mg/L, and a precipitation time of 20 min. Under these conditions, the COD content is 60 mg/L, the total phosphorus content is 0.26 mg/L, and the NH4-N content is 118.13 mg/L, COD and TP meet the secondary discharge standard of sewage. Using adding chlorine at break-point for the NH_4-N of treatment of garbage compression wastewater, when reacting under experimental conditions are as follow: pH 7, dosage (NaClO: N) 8, and aeration time of 30 min and the removal rate of NH_4-N reaches 89.70%, final NH_4-N content is reduced to 12.17 mg/L.

Keywords Garbage compression wastewater · Flocculation · Orthogonal test

L. Wang · P. Gao (✉) · H. Fu · Y. Dai · S. Liu
College of Environmental Science and Engineering, Xiamen University of Technology, Xiamen 361024, Fujian, China
e-mail: gaopf@xmut.edu.cn

C. Yuan et al. (eds.), *Proceedings of 4th International Conference on Resources and Environmental Research—ICRER 2022*, Environmental Science and Engineering,
https://doi.org/10.1007/978-3-031-31808-5_2

2.1 Introduction

Due to the huge foundation of waste generation, municipal solid waste has become an important environmental problem threatening the sustainable development of Chinese society (Ye 2011). The garbage transfer station is a garbage transfer station set up between the garbage origin and the treatment plant in order to reduce the transportation cost of the garbage removal process. Garbage compression wastewater is the pressure filtrate produced when garbage is compressed at the transfer station. If the wastewater pressure filtrate is improperly treated, it will directly threaten human health (Zhang and Zang 2018).

Garbage compressed wastewater has high COD concentration and high NH_4-N content (Jiao 2004), and produces non-centralized, small amount, and intermittent discharge. Traditional treatment methods are difficult to adapt to the basic requirements of in-situ, immediate, efficient, economical and reasonable garbage compression wastewater treatment (Yang and Lv 2005).

In the experimental study on wastewater treatment by flocculation (Zeng et al. 2007), the results show that the flocculation method can effectively treat wastewater (Malika and Sahab 2003) by analyzing the PFC dosage, PAM dosage, pH value and reaction time to determine the optimal treatment capacity. Relevant studies show that (Yang 2007), the flocculation effect of PFC-PAM composite flocculant on wastewater is obviously better than that of single flocculant. The dosage of reagent will have a significant impact on the flocculation effect (Ren et al. 2017). In the study of composite (PFC-PAM) flocculant, the modified PFC-PAM flocculant was found when the pH value was 7.00 and the dosage of flocculant was 5 mg/L. It has good settlement performance, which is better than PFC (Yan and Yang 2000). Combined with the characteristics of waste pressure filtrate, PFC-PAM method is optimized for the pretreatment of waste pressure filtrate.

For the treatment process of NH_4-N, we choose the breaking point chlorination method, which has higher removal efficiency, faster reaction speed, and lower requirements for temperature and equipment. The principle of the breakpoint chlorination method is to pass Cl_2 or ClO– into wastewater containing NH_4-N, and oxidize NH_4 + to N_2 to reduce the content of NH_4-N in the sewage. The removal of ammonia can be improved by adjusting the pH, reaction time and dosage as independent variables rate.

To sum up, the direct discharge of waste leachate without treatment is easy to cause secondary pollution, which can only be discharged into the municipal pipe network after treatment. In all experiments, due to the large number of variables, it is necessary to determine the optimal conditions for the treatment of pressurized filtrate. Therefore, all tests were carried out by orthogonal test, and this treatment process can also be applied to the treatment of black odorous micro water bodies.

Table 2.1 Comparison of water pollution components

	COD (mg/L)	NH_3-N (mg/L)	TP (mg/L)
Raw water quality	18,864	216.64	2.692
Irrigation water	150	/	5
Secondary emission Standard	100	20	1.0

2.2 Materials and Method

2.2.1 Sample Analysis

Go to Xiamen Lianban Garbage Transfer Station to retrieve the waste filter press waste water sample. The test data are shown in Table 2.1. It can be seen that compared with the secondary emission standard, there is a big gap between NH_4-N and COD, and the TP gap is not large, so COD and NH_4-N will be focused on.

2.2.2 Analytical Methods

The pH value measured with a pH meter (pHS-25, Shanghai INESA Scientific Instrument Co., Ltd., Shanghai, China). COD is determined by dichromate method (GB/T11914-89), NH_4-N is determined by Nessler's reagent photometric method (H535-2009), TP is determined by molybdate spectrophotometer (GB/T11893-89).

2.3 Results and Discussion

2.3.1 Orthogonal Test of Flocculation

According to other studies, it is found that flocculation sedimentation has a greater effect on the removal of various pollutants. First, a small pilot test was carried out. By adding PAM and PFC to the water sample at a ratio of 1:1, the water sample began to form a large number of flowers. After the precipitation is complete, use the supernatant for testing. It was found that COD was reduced. Experiments have shown that the addition of a certain proportion of PAM and PFC will reduce the concentration of COD. In order to find the best ratio of PAM and PFC and the best deposition time. We considered the pH, the amount of PAM and PFC and the precipitation time, and designed an orthogonal experiment, as shown in Table 2.2.

Table 2.2 Factors and standards of orthogonal test of flocculation

Level	pH	PAM (mL)	PFC (mL)	Precipitation time (min)
1	7.00	1.00	1.00	10.00
2	8.00	2.00	2.00	20.00
3	9.00	4.00	4.00	30.00
4	10.00	8.00	8.00	40.00

It can be seen from Table 2.3 that different pH, PFC and PAM dosages have a greater impact on COD. The removal rate of COD is between 91.94 and 99.70%. The effect of flocculation and sedimentation on COD removal is very significant. It can also be seen that this flocculation in the experiment, the removal rate of NH_4-N is not very obvious, basically about 50%.

Through orthogonal test, the best test ratio is obtained. The amount of PFC added is the main influencing factor. We concluded that the best treatment conditions for this water sample under the flocculation and sedimentation conditions were pH 7.00, PAM 2 mg/L, PFC 2 mg/L, and a precipitation time of 20 min. Under these conditions, the COD content is 60 mg/L, the total phosphorus content is 0.26 mg/L, and the NH_4-N content is 118.13 mg/L. Comparing with Xiamen's secondary sewage discharge standard, it is found that all the indicators except NH_4-N meet the discharge standard.

Table 2.3 Results of orthogonal test of flocculation

	pH	PAM (mL)	PFC (mL)	Precipitation Time (min)	COD removal rate (%)	TP removal rate (%)	NH_4-N removal rate (%)
Exp 1	7.00	1.00	1.00	10.00	99.58	92.31	59.72
Exp 2	7.00	2.00	2.00	20.00	99.67	90.32	45.47
Exp 3	7.00	4.00	4.00	30.00	99.58	18.38	32.46
Exp 4	7.00	8.00	8.00	40.00	97.46	19.12	27.92
Exp 5	8.00	1.00	1.00	10.00	99.70	88.58	62.92
Exp 6	8.00	2.00	2.00	20.00	98.73	78.23	58.71
Exp 7	8.00	4.00	4.00	30.00	95.76	9.60	10.71
Exp 8	8.00	8.00	8.00	40.00	96.61	39.08	47.02
Exp 9	9.00	1.00	1.00	10.00	99.15	9.52	42.72
Exp 10	9.00	2.00	2.00	20.00	98.30	18.30	29.14
Exp 11	9.00	4.00	4.00	30.00	99.15	77.99	34.00
Exp 12	9.00	8.00	8.00	40.00	91.94	78.57	51.99
Exp 13	10.00	1.00	1.00	10.00	98.73	7.04	22.41
Exp 14	10.00	2.00	2.00	20.00	93.64	31.79	35.32
Exp 15	10.00	4.00	4.00	30.00	99.58	92.97	56.41
Exp 16	10.00	8.00	8.00	40.00	94.91	68.30	62.14

Table 2.4 Factors and standards of orthogonal test of adding chlorine at break-point

Level	A pH	B dosing amount (NaClO:N)	C aeration time (min)
1	5.00	8.0	20
2	6.00	8.4	30
3	7.00	8.8	40

Therefore, the removal of COD and TP adopts the method of flocculation and sedimentation. This test also shows that a single flocculation and sedimentation cannot complete the purification treatment at one time.

2.3.2 *Treatment of NH_4-N by Adding Chlorine at Break-Point*

After flocculation precipitation, only the NH_4-N content is too high, we use the adding chlorine at break-point method to treat the NH_4-N.

According to the literature reviewed, when $m(Cl_2)$:m(N) is 8.2:1, the removal rate of NH_4-N is higher. When the amount of chlorine added is increasing, the removal efficiency of NH_4-N does not increase significantly; when the pH of the sewage is changed from 7 When it rises to 9, the removal rate of NH_4-N has a significant decreasing trend. Therefore, the breakpoint chlorination method is suitable to be carried out under neutral conditions; when the reaction time increases from 10 to 30 min, the content of NH_4-N significantly decreases. When the reaction time increases again, the removal rate of NH_4-N does not increase significantly. Therefore, we designed an orthogonal experiment based on the influencing factors of sewage pH, dosage ratio and reaction time, as shown in Table 2.4.

From the Table 2.5 of range analysis of the orthogonal test, it can be seen that the order of the three influencing factors on the removal efficiency of NH_4-N is dosing amount > pH > aeration time. The dosage has a significant effect on the removal of ammonia nitrogen. The optimal factor level combination is $A_3B_1C_3$, that is, the dosage is 8, the pH is 7, and the aeration time is 40 min, under the experimental conditions, the degradation efficiency of NH_4-N reached 89.70% and NH_4-N content is 12.17 mg/L. According to the optimal conditions, the formula of ammonia nitrogen remover is determined as follows: sodium hypochlorite 35% and terrazzo powder 65%.

2.3.3 *Cost Calculation*

PAM is 10 yuan/kg, PAC is 0.12 yuan/kg, industrial sodium hypochlorite is 18 yuan/kg, and terrazzo powder is 0.4 yuan/kg. According to the formula of ammonia nitrogen remover, the price of ammonia nitrogen remover is 6.6 yuan/kg. According

Table 2.5 Results of orthogonal test of adding chlorine at break-point

	A pH	B dosing amount	C aeration time	D	NH_4-N removal rate (%)
1	1	1	1	1	82.20
2	1	2	2	2	80.87
3	1	3	3	3	78.94
4	2	1	2	3	87.68
5	2	2	3	1	84.44
6	2	3	1	2	76.70
7	3	1	3	2	89.70
8	3	2	1	3	81.21
9	2	3	2	1	86.57
K_1	80.670	86.527	80.037	84.403	
K_2	82.940	82.173	85.040	82.423	
K_3	85.827	80.737	84.360	82.610	
R	5.157	5.790	5.003	1.980	

to the dosage of drugs, it only needs 33.02 yuan per ton of garbage pressurized wastewater to meet the discharge standard. Compared with other processes, the cost is lower and more suitable for popularization and use.

2.3.4 Processing Flow

After all the experimental processes, we have reduced the pollution standard of waste pressure filtration at Lianban waste transfer station in Xiamen to below the tertiary standard of sewage discharge in Xiamen. The general process of the experiment is shown in Fig. 2.1.

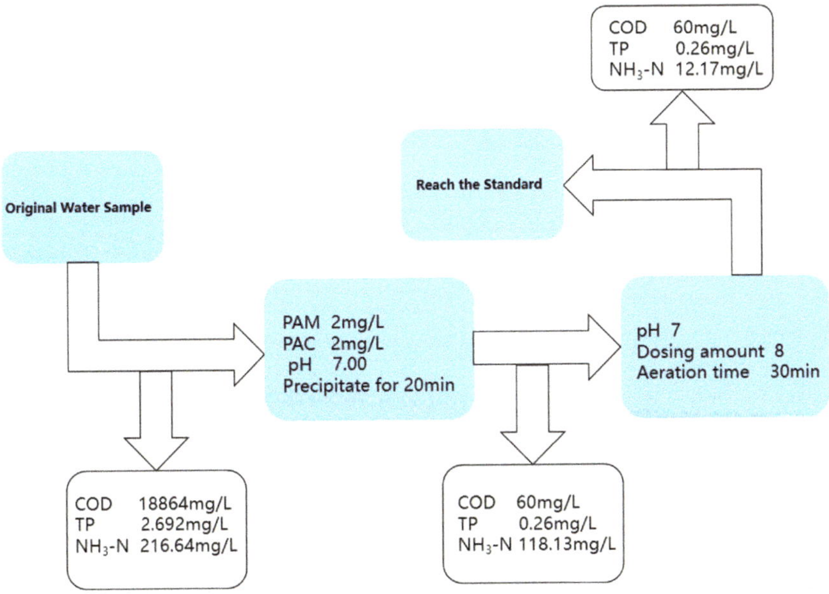

Fig. 2.1 Processing flow

2.4 Conclusion

(1) Choosing PAM-PFC flocculation sedimentation method can effectively remove COD and TP in waste compressed wastewater. When the experimental conditions are pH 7.00, PAM 2 mg/L, PFC 2 mg/L, and precipitation time 20 min. Under these conditions, the COD content is 60 mg/L and the total phosphorus content is 0.26 mg/L, both of which meet the emission standards.

(2) The adding chlorine at break-point method is selected for the degradation of ammonia nitrogen. Under the experimental conditions of pH 7, dosage (NaClO: N) was 8, and aeration time of 30 min, the content of ammonia nitrogen in compressed waste water can be effectively reduced to 12.17 mg/L.

(3) We adopt these two combined processes to realize the on-site solution of waste compressed wastewater, and the combined process is simple to operate, low in cost and easy to promote. The garbage wastewater can reach the secondary sewage discharge standard and irrigation water standard after the process is processed. It can be directly discharged into the municipal pipe network or irrigated farmland to realize the maximum utilization of water resources.

(4) The process can also be applied to the treatment of black odorous micro water bodies, and a series of measures such as point source pollution interception are adopted to prevent external pollution, continuously improve the water treatment effect and efficiency, and realize a virtuous cycle of the water environment in the basin.

Acknowledgements This research is financially supported by Fujian provincial industry-university-research collaborative innovation (2021Y4005), Fujian Science and Technology Guiding Project (2020Y0056), Fujian Engineering and Research Center of Rural Sewage Treatment and Water Safety, Xiamen University of Technology Postgraduate Science and Technology Innovation Program (YKICX2021154).

References

Jiao B (2004) Research on leachate control technology of municipal solid waste sanitary landfill site. Chongqing University

Malika PK, Sahab S (2003) Oxidation of direct dyes with hydrofen peroxide using ferrous ion as catalyst. Sep Purif Technol 31(3):241–250

Ren X, Liu H, Qin X, Zhang S (2017) Study on flocculation effect of composite flocculant (pac-pam) on pharmaceutical wastewater. J Heze Univ 39(02):77–81

Yan W, Yang H (2000) Preparation and properties of composite (pac-pam) flocculant. J Qinghai Teachers Coll 03:55–58

Yang D (2007) Physical and oxidative removal of organics during Fenton treatment of mature municipal landfill leachate. J Hazard Mater 146(1–2):334–340

Yang X, Lv C (2005) Study on environmental pollution of municipal solid waste compression station. Environ Pollut Prevent 27(8):618–620

Ye B (2011) Pilot study on biochemical regulation and rapid treatment technology of filter press in waste transfer station. Tsinghua University

Zeng Y, Yang C, Zhang J, Pu W (2007) Feasibility investigation of oily wastewater treatment by combination of zinc and PAM in coagulation flocculation. Hazard Mater 147(3):991–996

Zhang T, Zang X (2018) Research progress of landfill leachate treatment technology. Tianjin Chem Indus 32(06):1–4

Chapter 3
Comparative Study on the Removal Effect of Cyanobacteria by Two Algae Removal Schemes Using New Environmental Protection Oxidant

J. C. Tu, M. L. Zhu, Z. Lin, H. Jiang, and Y. L. Han

Abstract In recent years, the rapid development of society has led to serious water pollution, and many water bodies have appeared eutrophication and water bloom in varying degrees. In this study, two new, green and environmentally friendly super oxidants, namely sodium percarbonate and potassium bisulfate, are used for the algae waters in ponds where serious cyanobacteria bloom occurs, and potassium ferrate with functions of oxidation, adsorption, flocculation and precipitation is used in combination. Two algal removal schemes: scheme 1 uses sodium percarbonate alone (dosage 4.5 mg/l), and scheme 2 uses potassium bisulfate alone (dosage 4.5 mg/l) + potassium ferrate (dosage 4.5 mg/l) to carry out the algal removal experiments with 123 h. Through visual observation and analysis of the change process of pH and water temperature of algae water and the change of chlorophyll a content (cyanobacteria) at the beginning and end of the experiment, the following conclusions are obtained: the algae removal effect of scheme 1 is significantly better than that of scheme 2. At the end of the experiment, there are basically no cyanobacteria and dead cyanobacteria in the experimental barrel of scheme 1. The removal effect of sodium percarbonate in scheme 1 on cyanobacteria began to take effect after the experiment lasted 34–58 h, and the removal rate of chlorophyll a (cyanobacteria) was 86.22% at 123 h. The removal effect of potassium bisulfate (PMS) on cyanobacteria began to take effect in (16–34 h), but the removal effect of chlorophyll a (cyanobacteria) would rebound within 24 h and could not be sustained. Under the synergistic effect of potassium ferrate, the removal rate of chlorophyll a (cyanobacteria) was 76.71% at 123 h. The research results can provide a technical reference for the control of cyanobacteria bloom in the future.

J. C. Tu · M. L. Zhu (✉) · Z. Lin · H. Jiang · Y. L. Han
School of Environmental Science and Engineering, Xiamen University of Technology, Xiamen 361024, China
e-mail: 292890310@qq.com

The Key Laboratory of Water Resources Utilization and Protection of Xiamen, Xiamen 361024, China

C. Yuan et al. (eds.), *Proceedings of 4th International Conference on Resources and Environmental Research—ICRER 2022*, Environmental Science and Engineering,
https://doi.org/10.1007/978-3-031-31808-5_3

Keywords Remove cyanobacteria · Comparison of oxidants · Sodium percarbonate

3.1 Introduction

With the rapid development of China's economy, the phenomenon of industrial and agricultural production wastewater and domestic sewage being discharged in disorder is becoming more and more serious. A large number of pollutants are discharged into rivers, lakes and other water bodies, resulting in eutrophication, black odor and water bloom. According to statistics, more than half of the natural water bodies in China have different degrees of eutrophication, which deserves our attention (Freedma 2002; Qin et al. 2013). However, the eutrophication of the water body is often accompanied by cyanobacteria bloom, resulting in the death of a large number of fish in the water body and the production of algal toxins (MCS), heteroolfactory substances and other algal organic matter (AOM), which seriously endanger the water environment (Li et al. 2012, 2008a).

At present, the control methods of cyanobacteria bloom mainly include physical method, chemical method and biological method. Among them, physical methods such as physical salvage and other technologies, although the treatment effect is good, often have high treatment costs and easy recurrence, and are difficult to apply to practical projects; and biological methods such as aquatic higher plants control algae, aquatic animals control algae and other technologies also have slow effects. It is difficult to implement and may cause biological disasters due to the introduction of alien species; while traditional chemical methods such as copper sulfate algae removal will bring heavy metal pollution, and the killed algae will also produce highly toxic algal toxins, bringing about secondary pollution problems that cannot be ignored (Fang et al. 2020).

In recent years, some new environmentally friendly oxidants such as persulfate and percarbonate have begun to attract researchers' attention (Wang 2021; Liao et al. 2022). For percarbonate, Sinha, A. K. et al. took the pond water with total chlorophyll a of 1000 ug/L as the object and found that percarbonate can effectively remove algae at 4.0 mg/L, and the removal efficiency is about 81.8% (Sinha and Green 2018). For persulfate, it is divided into two types: monopersulfate and peroxodisulfate, among which potassium hydrogen persulfate (PMS) belongs to monopersulfate and sodium persulfate belongs to peroxodisulfate (Xu 2018). Tsiarta et al. (2017) took the artificially cultured $3 * 10^6$ cells/mL Microcystis aeruginosa as the research object and found that the removal rate of algae by PMS was about 31.68%, which was greater than that of sodium persulfate, which was about 4.2%, and the removal rate of algae by PMS was more Suitable for algae removal. PMS mainly degrades pollutants in water by generating free radicals such as ·OH or $SO_4^{-}\cdot$ through a chain reaction (Liu et al. 2021). In order to increase the removal rate, PMS needs to generate more chain reactions. Some scholars have pointed out that PMS can be catalyzed to generate more oxidative groups through UV irradiation, transition metal ions and oxidants synergistically (Matilainen and Sillanpää 2010). For example, Wang et al. (2021) took the artificially cultured 10^6 cells/mL Microcystis aeruginosa as the research object and found that the removal rate of PMS to algae was as high as 94.66% under the synergistic effect of $FeSO_4$. There are relatively few studies on the synergistic catalysis of PMS by oxidants to generate chain reactions. Taking $2.39*10^9$/L lake water as the object, Li et al. (2008b) found that the removal rate of algae could reach 72% under the action of 0.26 mg/L ferrate. It is a synergistic oxidant and has a good algae removal effect. At present, there are few experimental studies on the combined use of PMS and potassium ferrate for algae removal, and a large number of studies focus on the application of PMS, and there are few studies on percarbonate; especially compared with this the research on the removal of cyanobacteria by the two new oxidants is rare. At the same time, the existing research lacks to investigate whether the oxidant will harm or even cause the death of aquatic animals such as fish fry during the algae removal process. At the same time, the existing research lacks the ability to investigate whether the oxidant will cause damage or even death to aquatic animals such as fish fry in the process of algae removal. Because the risk is unclear, it is difficult to directly apply the existing research results to the treatment of blue-green algae bloom in small and micro water bodies where various aquatic animals grow.

To this end, this paper intends to take the actual pond algal water with severe cyanobacterial blooms and high cyanobacterial content as the research object, using sodium percarbonate and potassium persulfate, two new environmentally friendly oxidants, and using them in combination with oxidation, adsorption The synergistic effect of potassium ferrate, flocculation and precipitation, etc., has designed two algae removal technical schemes, and carried out a comparative study on the removal effect of these two schemes on cyanobacteria and the impact on the survival of small goldfish, in order to control cyanobacteria blooms in the future The small and micro water bodies provide a good technical solution for algae control.

3.2 Experimental Materials and Methods

3.2.1 Experimental Water

The experimental water is taken from a local pond where serious cyanobacteria bloom occurs (see Fig. 3.1). The initial pH of the experimental water and the chlorophyll a content (cyanobacteria), chlorophyll a content (green algae) and chlorophyll a content (dinoflagellate/diatom) reflecting the concentration of algae are shown in Table 3.1. It can be seen from Table 3.1 that the content of chlorophyllin a (cyanobacteria) in this pond is much higher than that of other algae. It can be seen that the bloom in this pond belongs to cyanobacteria bloom. For this reason, the follow-up experiments only monitored the chlorophyll content a (cyanobacteria).

Fig. 3.1 Pond with experimental water

Table 3.1 Initial chlorophyll a content, pH and water temperature of experimental water

Water quality index	Chlorophyll a (cyanobacteria) (ug/L)	Chlorophyll a (green algae) (ug/L)	Chlorophyll a (diatoms) (ug/L)	pH	Water temperature °C
Detect the value	352.33	0	28.63	7.75	21.1

3.2.2 *Experimental Scheme*

Two new environmental protection oxidants, namely sodium percarbonate (supplier: Wuhan lanabai pharmaceutical and Chemical Co., Ltd.) and potassium bisulfate (supplier: Jiangsu pharmaceutical) were used in this study. As well as a flocculating precipitant used in combination with potassium bisulfate, i.e. potassium ferrate (supplier: Jiangsu pharmaceutical) with synergistic effects of oxidation, adsorption, flocculation and precipitation, two algal removal experimental schemes are designed as shown in Table 3.2. Scheme 1: use sodium percarbonate new oxidant alone to remove algae; In scheme 2, in order to enhance the algal removal effect of potassium bisulfate, after 50 h of algal removal by potassium bisulfate, potassium ferrate is further added to enhance the oxidation of algae and absorb the killed algae in the flocculated water.

The experimental device is 2 buckets, each bucket contains 15 L of pond algal water, and each bucket is filled with a small goldfish which is used for water quality warning in the water source and is very sensitive to water quality. Its length is about 3.5 cm. The experimental water bucket is placed at the corner of the roof of the laboratory building of our institute to prevent the water temperature from rising too high due to direct sunlight for a long time. The experiment lasted for 123 h from 20:00 on May 12 to 23:00 on May 17. Sodium percarbonate and potassium persulfate were added at 21:00 on May 12, 1 h after the experiment began, and potassium ferrate was added at 23:00 on May 14, 50 h after the experiment began. It should be specially

Table 3.2 Details of two algae removal experimental schemes

Scheme	Experimental barrel volume (L)	Drug name (dosage)	Drug input mode	Fish feeding in the experimental barrel
1	15	Sodium percarbonate (4.5 mg/l)	One hour after the start of the experiment, 15 * 4.5 = 67.5 mg sodium percarbonate was dissolved in water at a ratio of 1:200, and then evenly sprinkled into the water surface of the experimental bucket	One small goldfish with a length of about 3.5 cm
2	15	Potassium bisulfate (4.5 mg/l) + potassium ferrate (4.5 mg/l)	One hour after the start of the experiment, 15 * 4.5 = 67.5 mg potassium persulfate was dissolved in water at a ratio of 1:200, and then evenly sprinkled into the water surface of the experimental bucket. After that, 50 h after the start of the experiment, 15 * 4.5 = 67.5 mg potassium ferrate was dissolved in water at a ratio of 1:200 and then evenly sprinkled into the water surface of the experimental water bucket at one time	One small goldfish with a length of about 3.5 cm

pointed out that the reason why the drug was not added immediately at the beginning of the experiment, but was added after 1 h was that there was a certain impact force when the algal water was stirred evenly and poured into the experimental barrel at the beginning of the experiment. It was necessary to give the algae a short period of time to mitigate the impact of the impact force and adapt to the new growth environment. See Table 3.2 for details of the two algae removal experimental schemes.

3.2.3 *Water Quality Monitoring Indicators and Detection Methods*

There are three water quality indicators monitored in this study, namely, chlorophyll a (cyanobacteria), pH and water temperature. The detection methods of various water quality indexes are shown in Table 3.3.

Due to the phototaxis of algae, a large number of blue-green algae float on the surface of the bucket during the day, while the algae content under the water surface is very low. Therefore, it is difficult to accurately reflect the chlorophyll a content in the water body, whether it is the surface sampling or the bottom sampling. In order to

Table 3.3 Test methods and instruments for water quality indicators

Water quality index	Use instrument	Test method
Total chlorophyll a (Cyanobacteria)	Phytoplankton classification fluorometer	Spectrophotometry (HJ 897-2017)
pH	Portable pH meter	Glass electrode method (GB/T6920-1986)
Water temperature	Portable pH meter	Thermistor sensor method (DB22/T 3102-2020)

correctly reflect the algae content in the experimental water, this study only slightly stirred the experimental water at the beginning and end of the experiment. After the mixing was uniform, the samples were taken, and then the chlorophyll a content in the water samples was detected. During the experiment, in order to reduce the impact of agitation on the growth of algae, no sampling and detection of chlorophyll a content were carried out. Instead, a portable pH meter was used to monitor the pH and water temperature in the water body 5 cm below the water surface. The monitoring time points of pH and water temperature were at the beginning and end of the experiment, and 5 fixed time points every day, i.e. 6:00, 12:00, 16:00, 18:00 and 24:00 (but the last day of the experiment, i.e. May 17, was only monitored at 7:00, 15:00 and 23:00).

As we all know, the algae will carry out photosynthesis during the day and absorb CO_2 in the water to convert it into O_2. Since CO_2 is dissolved in water and becomes acidic, the acidity of the water body will decrease after the CO_2 in the water is absorbed by the algae and reduced, that is, the pH will increase; In the evening, the algae will breathe, absorb O_2 in the water and convert it into CO_2, resulting in an increase in the acidity of the water, that is, a decrease in pH. Therefore, the pH value of the water body will show a law of day and night fluctuation, and the more the algae content, the greater the diurnal fluctuation range of the pH value. Su (2016) Studied the changes of pH of Microcystis aeruginosa and Chlorella under different light intensities and water temperatures, and found that the pH of water environment and the density of algae cells were positively correlated. It can be seen that by observing the change of pH, it is expected to reveal the change trend of algae concentration during the algae removal experiment.

3.3 Results and Discussion

3.3.1 *Qualitative Observations*

Visual Observation Results of Algae Changes. The photos of the algal concentration change process of the experimental algal water of schemes 1 and 2 during the whole experiment are shown in Figs. 3.2 and 3.3 respectively.

It can be seen from Figs. 3.2 and 3.3 that in scheme 1, after 34 h of the experiment, the concentration of algae has not changed significantly, but by 58 h, the concentration of algae has decreased significantly. At the end of the experiment (123 h), the algae has basically disappeared, and even the algal corpse has been completely oxidized and disappeared; In scheme 2, after 34 h of the experiment, the thick layer of blue-green algae floating on the water surface became thinner. However, after 58 h of the experiment, the concentration of algae did not significantly decrease, which was in sharp contrast with scheme 1. At the end of the experiment in scheme 2, although there was no layer of blue-green algae on the water surface and a large number of blue-green algae had died, a large number of algal corpses were piled up at the bottom of the barrel and were not completely oxidized and disappeared. According to the visual observation results, scheme 1 (sodium percarbonate) has better algal removal effect than scheme 2 (potassium bisulfate + potassium ferrate).

Visual Observation Results of Survival State of Small Goldfish. During the experiment, when sodium carbonate was added in scheme 1, the small goldfish became active and swam up and down in the bucket. After several hours, the agitation disappeared and returned to normal. Until the end of the experiment, the small

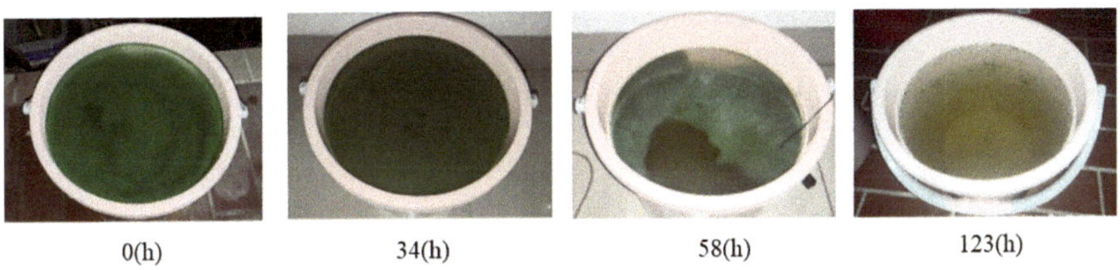

Fig. 3.2 Visual observation results of algae concentration change in scheme 1

Fig. 3.3 Visual observation results of algae concentration change in scheme 2

goldfish were alive. In scheme 2, the small goldfish did not have any abnormal reaction from the beginning to the end of the experiment, and the small goldfish survived normally.

3.3.2 Analysis of Quantitative Monitoring Results

Changes of Water Temperature with Time in Two Algal Removal Experimental Schemes. During the experiment, the water temperature change process lines of schemes 1 and 2 are shown in Fig. 3.4 respectively. It can be seen from Fig. 3.4 that the water temperature changes of the two schemes show similar diurnal fluctuation changes with a cycle of 24 h. The water temperature variation ranges of scheme 1 and scheme 2 are the same in the first cycle and the second cycle, which are 19.5–26.3 °C and 23.2–24.6 °C respectively. In the third cycle, the water temperature change in scheme 1 is 20.8–27.8 °C, and that in scheme 2 is 21.5–27.8 °C. In the fourth cycle, the water temperature change of scheme 1 is 20.8–26.3 °C, and that of scheme 2 is 21.5–26.9 °C. In the fifth cycle, the water temperature change of scheme 1 is 20.8–26.1 °C, and that of scheme 2 is 21.5–26.2 °C. The variation range of water temperature in scheme 1 and scheme 2 is approximately the same, and the overall variation range is 19.5–27.8 °C.

Change of pH Value with Time of Two Algal Removal Experimental Schemes. As can be seen from Fig. 3.5.

The pH value change curves of schemes 1 and 2 have a diurnal fluctuation law with a cycle of 24 h. Each cycle includes a rising period from trough to peak and a falling period from peak to trough. The peaks of pH mostly appear at 16:00 p.m. and the troughs mostly appear at 6:00 a.m.

In scheme 1, the amplitudes of the rising phase and the falling phase in the first cycle (the experiment lasted 10–34 h) were 2.61 and 2.25 respectively, and

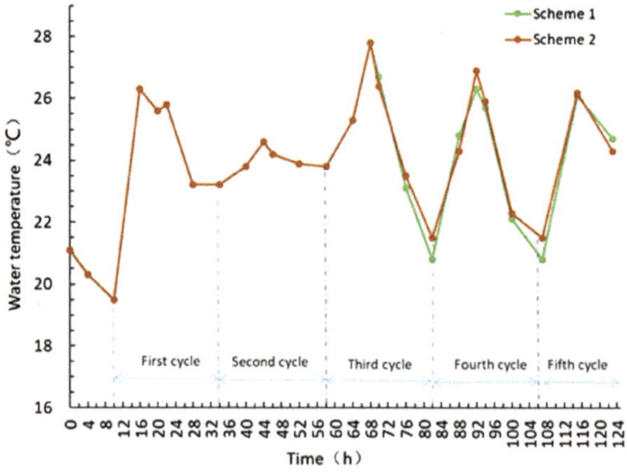

Fig. 3.4 Change process of water temperature with time for two algae removal schemes

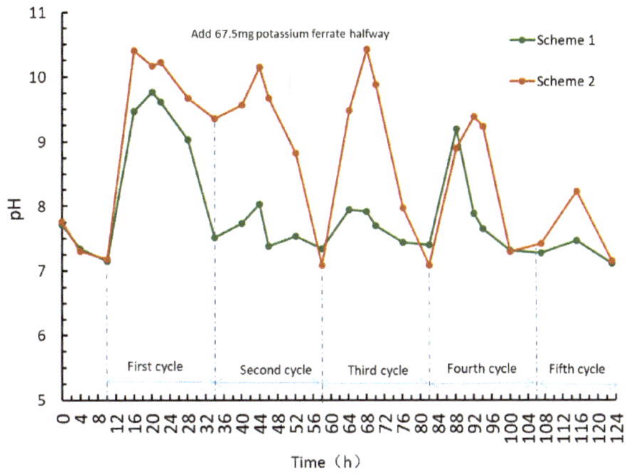

Fig. 3.5 Change process of pH value with time of two algae removal schemes

the amplitude of the falling phase was slightly smaller than that of the rising phase; And from the second cycle (34–58 h), the diurnal amplitude of pH value significantly decreased to only 0.51 and 0.65, and with the passage of time, the amplitude further attenuated, and the final amplitude (the amplitude in the rising period of the fifth cycle) was only 0.2; It can be seen that sodium percarbonate has a good removal effect on cyanobacteria, and it can show a very significant removal effect after 34–58 h.

The amplitude of scheme 2 in the first cycle (the experiment lasted 10–34 h) was 3.22 in the rising phase, and the amplitude in the falling phase was significantly smaller, only 1.05, which indicates that under the oxidation of potassium persulfate, the function of many algae was damaged in the first cycle, and normal respiration and CO_2 production were not possible, resulting in the pH value could not be reduced from night to early morning. After that, the peak value in the rising period of the second cycle (34–58 h) is also slightly lower than the peak value in the first cycle, but the valley in the falling period returns to the initial low again, and the amplitude in the third cycle (58–82 h) completely returns to the initial state. The reason may be that the algae realize resistance to the oxidation of Potassium Bisulfate under the action of SOD enzyme (superoxide dismutase) automatically generated after the environment is stressed, The research of Zhang et al. (2011) Showed that the algae will automatically produce SOD enzymes to enhance the antioxidant system and thus enhance the resistance to adversity. Subsequently, under the synergistic effect of further oxidation, adsorption and flocculation of potassium ferrate added at the 50th hour, the pH amplitude of the algae began to decay in the fourth cycle, and the amplitude of the fifth cycle further attenuated to 0.81.

To sum up, although potassium bisulfate can take effect earlier than sodium percarbonate, and significant pH amplitude decreases in the falling period of the first cycle, while carbon percarbonate does not show significant pH amplitude attenuation until the rising period of the second cycle, the algal water added with potassium bisulfate will rebound, and the final pH amplitude is also larger than that of sodium percarbonate. It can be seen that scheme 1 (sodium percarbonate) has better algal removal

Table 3.4 Comparison of chlorophyll a content (cyanobacteria) removal rate of two algae removal schemes

Algae removal scheme	Initial chlorophyll a content (Cyanobacteria) (ug/L)	Chlorophyll a content at the end of the experiment (Cyanobacteria) (ug/L)	Chlorophyll a content (Cyanobacteria) removal rate (%)
1	352.33	48.52	86.22
2	352.33	82.03	76.71

performance than scheme 2 (potassium bisulfate + potassium ferrate), which is consistent with the visual observation results.

Removal Effect of Chlorophyll a Content (Cyanobacteria) in Two Experimental Schemes. The initial chlorophyll a content (cyanobacteria), the chlorophyll a content (cyanobacteria) at the end of the experiment and the removal rate of chlorophyll a content (cyanobacteria) of the experimental algae water in schemes 1 and 2 are shown in Table 3.4.

It can be seen from Table 3.4 that the removal rates of chlorophyll a content (cyanobacteria) in scheme 1 and scheme 2 are 86.22 and 76.71%, respectively. It can be seen that the algae removal effect of scheme 1 is significantly better than that of scheme 2. This is consistent with the visual observation results and the algal removal effect obtained from the above analysis based on the change of pH.

3.4 Conclusion

In this study, two new, green and environmentally friendly super oxidants, namely sodium percarbonate and potassium bisulfate, and potassium ferrate with the functions of oxidation, adsorption, flocculation and precipitation, were used to remove algae from the algae water of a local pond where serious cyanobacteria bloom occurred. Two algae removal schemes were designed, in which sodium percarbonate (dosage 4.5 mg/l) was used alone in scheme 1, Scheme 2 uses potassium bisulfate (dosage 4.5 mg/l) + potassium ferrate (dosage 4.5 mg/l). A 123 h algae removal experiment was carried out for these two algae removal schemes. By analyzing the change process of pH and water temperature of algae water and the change of chlorophyll a content (cyanobacteria) at the beginning and end of the experiment, the following conclusions were obtained:

(1) During the experiment, the water temperature of the two schemes showed the same diurnal change law with a cycle of 24 h, and the water temperature changes were basically the same, with a range of 19.5–27.8. The pH value also shows a diurnal fluctuation law with a cycle of 24 h, and the larger the algae content, the larger the amplitude of the fluctuation change, and vice versa. Throughout the experiment, the fluctuation range of pH value in scheme I is 7.11–9.76, and that in scheme II is 7.08–10.42.

(2) In scheme 1 with sodium percarbonate, the amplitude of pH fluctuation changes significantly attenuated in the second cycle (34–58 h), and the amplitude decreased from 2.61 to 0.51 at the beginning. After that, the amplitude steadily decreased with time, and finally decreased to 0.2. It can be seen that the algal removal effect of sodium percarbonate can be effective in the second cycle (34–58 h). The visual observation results also verify the analysis results. The photo of the experimental algal water lasting for 58 h shows that the algal concentration is significantly lower than the initial concentration.

(3) In scheme 2 of potassium bisulfate + potassium ferrate, the amplitude of pH value has significantly decreased in the falling period of the first cycle (10–34 h), and the amplitude has decreased from 3.22 to 1.05. However, in the following second and third periods (34–58 h and 58–82 h), the amplitude returned to the initial value of about 3.34. This indicates that the algal removal effect of potassium persulfate can be effective in the second half of the first cycle (10–34 h), but the algal removal effect cannot be sustained and bounces back within 24 h, which may be due to the damage of the algae that secrete SOD enzymes on their own and have resistance to the oxidizing environment of potassium bisulfate. After that, under the synergistic effect of oxidation, adsorption and flocculation of potassium ferrate added subsequently, the pH amplitude began to decay in the fourth cycle and decreased to 0.81 in the fifth cycle.

(4) After 123 h of experiment, the removal rate of chlorophyll a content (cyanobacteria) in scheme 1 was as high as 86.22%, while that in scheme 2 was 76.71%. The removal rate of chlorophyll a content (cyanobacteria) in scheme 1 is 9.51% higher than that in scheme 2. In addition, it can also be seen from the visual observation results that at the end of the experiment, there was basically no algal corpse at the bottom of the experimental barrel in scheme 1, and almost all algal corpses were oxidized and decomposed. However, there are a large number of algal corpses at the bottom of the experimental barrel in scheme 2. It can be seen that the algal removal effect of scheme 1 using sodium percarbonate is significantly better than that of scheme 2 using potassium persulfate + potassium ferrate.

(5) By the end of the experiment, the small goldfish were all alive. Although the small goldfish in scheme 1 swam back and forth actively after the sodium bicarbonate was added, it did not affect their survival. Therefore, scheme 1 is preferentially recommended for the treatment of cyanobacteria bloom in small and micro water bodies.

Acknowledgements This work was financially supported by "Fujian Provincial Science and Technology Project (2023|01010022)".

References

Fang YB, Wang CY, Tang WW et al (2020) Comparison of advantages and disadvantages of algae removal technology, application status and new technology progress. Ind Water Treat 40(9):1–6

Freedma NB (2002) Environmental ecology. Academic Press, San Diego

Li L, Gao N, Deng Y et al (2012) Characterization of intracellular & extracellular algae organic matters (AOM) of Microcystic aeruginosa and formation of AOM-associated disinfection byproducts and odor & taste compounds. Water Res 46(4):1233–1240

Li L, Gao NY, Yin DD et al (2008a) Treatment processes in waterworks for removal of algae and algae toxins from raw Water. China Water Wastewater 24(6):20–24

Li L, He ZG, Wang TP et al (2008b) Study on algae removal by ferrate. Safety Environ Eng 15(4):51–54, 76

Liao B, Xu W, Ye QY (2022) A review of activated percarbonate and peroxymonocarbonate in the field of water treatment. Chem Ind Eng Progress 41(6):3235–3248

Liu YC, Yang B, Li QM et al (2021) Effects and mechanism of Cl-and pH on organic matter removal in salt-containing wastewater treatment by advanced oxidation processes. Chinese J Environ Eng 15(5):1487–1499

Matilainen A, Sillanpää M (2010) Removal of natural organic matter from drinking water by advanced oxidation processes. Chemosphere 80:351–365

Qin BQ, Gao G, Zhu GW (2013) Lake eutrophication and its ecosystem response. Chin Sci Bull 58(10):855–864

Sinha AK, Green WR (2018) Mitigating cyanobacterial blooms and cyanotoxins in hypereutrophic ponds following the application of a granular hydrogen peroxide-based algaecide. Arkansas Bull Water Res 28–38

Su FW (2016) Composition of phytoplankton community structure in Dafeng saline alkali shrimp pond and its impact on water environment pH. Shanghai Ocean University

Tsiarta N, Schuurmans J, Matthijs H et al (2017) Mode of action of hydrogen peroxide, peroxymonosulfate and persulfate on microcystis aeruginosa strain pcc 7806. In: Proceedings of the 15th international conference on environmental science and technology, Rhodos, Greece, p. 31

Wang L, Wei Q, Ma XM et al (2021) Removal of Microcystis aeruginosa by synergy of ferrous sulfate and potassium hydrogen persulfate. Chinese J Environ Eng 15(11):3572–3580

Wang SP (2021) Study on bisphenol a degradation in water by new magnetic biochar activated monopersulfate. Hefei University of Technology

Xu N (2018) Experimental study on enhanced coagulation of high algae water with potassium bisulfate. Shandong University of Architecture and Architecture

Zhang RF, Tang DS, Liu F (2011) Alage antioxidant system and its influence on stress. Environ Sci Manag 36(12):21–25

Chapter 4
Effects of Polyethylene Microplastics and Natural Sands on the Dispersion of Spilled Oil in the Marine Environment

Xin Ping Yu, Huan Gao, Ya Ya An, Zhi Xin Qi, and De Qi Xiong

Abstract The marine oil transportation and the use of plastic products have been increasing with the development of industrialization, resulting in frequent marine oil spill accidents and a great number of microplastics entering the ocean. The existence of microplastics in the surface seawater might have a certain impact on the migration and transformation of spilled oil. This paper explored the adsorption and aggregation effects of different concentrations of microplastics and natural sands on oil droplets by conical flask oscillation experiments. The results show that the concentration of oil-suspended particulate aggregates (OSA) formed by the spilled oil and microplastics is lower than that of OSA formed with natural sands, where the OSA concentration gradually decreases with the increase of microplastics concentration, ranging from 2.25–6.52%, while it gradually increases with the concentration of natural sands, up to 44.37%, and the particle size of OSA formed is larger in the former than in the latter. The results show that the concentration of oil-suspended particulate aggregates (OSA) formed by the spilled oil with microplastics is lower than that of OSA formed with natural sands.

Keywords Microplastics · Oil-suspended particulate matter aggregate · Natural sands · Oil spill

4.1 Introduction

With the introduction of plastics into all aspects of human life and the increasing amount of plastic waste disposal, the number of plastic debris released into nature has increased. It is estimated that approximately 8 million tons of plastic are released into the ocean each year (Wang et al. 2020). Thompson (Thompson et al. 2004) found that plastic was archived in plankton samples as early as the 1960s and that the

X. P. Yu · H. Gao · Y. Y. An · Z. X. Qi · D. Q. Xiong (✉)
College of Environmental Science and Engineering, Dalian Maritime University, Dalian 116026, China
e-mail: xiongdq@dlmu.edu.cn

C. Yuan et al. (eds.), *Proceedings of 4th International Conference on Resources and Environmental Research—ICRER 2022*, Environmental Science and Engineering,
https://doi.org/10.1007/978-3-031-31808-5_4

amount of plastic has increased significantly over time. Usually, marine plastic litter is divided into two main categories: macro plastics, with a particle size of 45 mm, and micro-plastics, which are plastic fragments, films and granules with a particle size of < 5 mm (Moore 2008; Browne et al. 2011; Eriksen et al. 2014).

Through years of research, microplastics have been found to be chemically hazardous in some cases, either due to their own added chemicals or the absorption of other toxic and hazardous substances from the environment (Mai et al. 2018), especially when long-term exposure in the environment increases the ability of microplastics to adsorb pollutants (Fisner et al. 2013), which can cause serious harm to aquatic organisms and the environment. When oil spills occur in the ocean, microplastics in seawater have the ability to adsorb polycyclic aromatic hydrocarbons (PAHs). Shan (Shan et al. 2020) simulated the behavior of adsorption and desorption of crude oil by MPs in the fish gut and found that the presence of microplastics increased the oil content of the aqueous phase of the gut fluid due to the presence of microplastics, while the temperature also affected the adsorption and desorption to some extent. The effects of MP concentration, MP aging, and dispersant-to-oil volume ratios (DORs) on oil dispersion were studied, and it was found that the aged microplastics were easier to disperse crude oil, while the competition of dispersants affected the dispersion of crude oil (Yang et al. 2021a). Natural sands, which are found everywhere in the ocean, also aggregates with spilled oil and the formed oil-suspended particulate matter aggregate (OSA) settles to the seafloor and accumulates heavier hydrocarbons in benthic organisms (Menon and Menon 1999), while OSA suspended in the water column is ingested by fish and facilitates the uptake and digestion of PAHs by organisms in the presence of particulate matter (Zhai et al. 2018).

Both microplastics and spilled oil can be extremely harmful to marine organisms and human beings. Based on the action of sea breaking waves and wind forcing, microplastics or natural sands will be adsorbed and aggregated with crude oil in the water column, forming suspended or settled oil-suspended particulate matter aggregates, or OSA for short. which has a certain impact on the ecological environment. The study of the aggregation behavior of microplastics and natural sands with spilled oil can provide some theoretical basis for the subsequent treatment of offshore oil spill removal and recovery (Yang et al. 2021a; Yu et al. 2021; Qi et al. 2021). In this study, the effect of microplastics and natural sands concentration on the amount of dispersed oil in water was investigated, and the particle size of the aggregates was measured to further discover the formation pattern.

4.2 Materials and Methods

4.2.1 Experimental Materials

The seawater used in this study was taken from the Xinghai Bay Baths of Dalian, China, Bay Baths, and the returned seawater was left for a period of time to precipitate coarse material and filtered with a 0.45 μm aqueous microporous filter membrane. The experimental oil was Roncador crude oil with a density of 0.94 g/cm^3, kinematic viscosity of 1248.78 mm^2/s and asphaltene content of 4.41%. In this study, natural sands and polyethylene (PE) were selected as suspended particulate matter in seawater. The natural sands was taken from the intertidal zone of Xiajiahezi Beach in Dalian, China, and the collected natural sands were pretreated by washing with hexane to remove the petroleum hydrocarbons and then dried, grounded and sieved with a 1000 mesh sieve; the PE powder was purchased with a particle size of about 1000 mesh. Both were stored in sealed containers for subsequent use.

The main chemical agents purchased in this experiment were n-hexane, anhydrous ethanol, concentrated sulfuric acid and carbon tetrachloride, all of which were purchased from Tianjin Comio Chemical Reagent Co. The experimental equipment included UV spectrophotometer (EPOCH2), in situ scattering laser particle size analyzer (LISST-100X), inverted fluorescence microscope (Olympus IX73), electronic balance (JJ6000Y), CNC ultrasonic cleaner (KQ5200DE), blast drying oven (DHG-9035A), diaphragm vacuum pump (GM- 0.33A) and full-temperature culture shaker (HZQ-QA).

4.2.2 Experimental Method

In this study, 360 mL of filtered seawater was added to a 500 mL conical flask, mixed with natural sands/microplastics, and then 0.3 mL of Roncador crude oil was added dropwise and shaken in a constant temperature shaker at 160 r/min to simulate the sea conditions at the time of the oil spill. The suspension in the conical flask was removed at five time points of 10, 30, 60, 120 and 240 min, and part of the suspension was added into a parting funnel with hexane, and the concentration of dispersed oil in the water was determined after extraction with hexane.

4.2.3 Droplet Size and Dispersion Effect

The oil droplet size is measured by an in-situ scattering laser particle size analyzer (LISST-100X), which is suitable for measuring the particle size of suspensions between 2.5 and 500 μm. The removed suspension was transferred to the particle size analyzer measurement chamber and the oil-suspension particle size aggregates were scattered at a small angle by a diode laser emitter, and the volume median particle size volume median diameter (*VMD*) of the aggregates size was rewritten according to the formula as:

$$VMD = \sum_{i=1}^{32} \frac{V_i}{C_{TV}} D_i \quad (4.1)$$

$$C_{TV} = \sum_{i=1}^{32} V_i \quad (4.2)$$

The dispersion effect was then calculated from the data obtained by UV spectrophotometer at a wavelength of 225 nm. Using hexane as the blank sample, the mass of dispersed oil was calculated as:

$$M_D = \frac{1}{3} \sum_{i=1}^{3} \frac{\omega \times C_i}{V_s} \times V_{sea} \times 1000 \quad (4.3)$$

where M_D denotes the mass of dispersed oil in mg; V_{sea} is the volume of seawater added to the experimental conical flask in 360 mL; C_i is the concentration of oil detected for each sample in mg/L; V_s is the volume of suspension removed for each sample in 25 mL; ω denotes the corresponding dilution multiple of the sample.

The oil dispersion efficiency (*ODE*) of the surface oil slick into water is related to the total amount of oil added to the conical flask and the amount of dispersed oil (Yang et al. 2021a), i.e., the ratio of the amount of oil dispersed in the water column to the total amount of oil added:

$$ODE = \frac{M_D}{M_T} \times 100\% \quad (4.4)$$

where M_D indicates the mass of dispersed oil in mg; M_T is the total amount of oil added to the water column in mg. All the above experiments were run in triplicate.

4.3 Results and Discussion

4.3.1 Morphology of OSA Formation

The OSA morphology of oil with microplastics and natural sands in the steady state, respectively, was first observed using inverted fluorescence microscopy (错误!未找到引用源。). It can be seen that there are obvious differences in the OSA morphology formed by different types of suspended particulate matter. The microplastics are completely wrapped by the crude oil and gathered at the junction

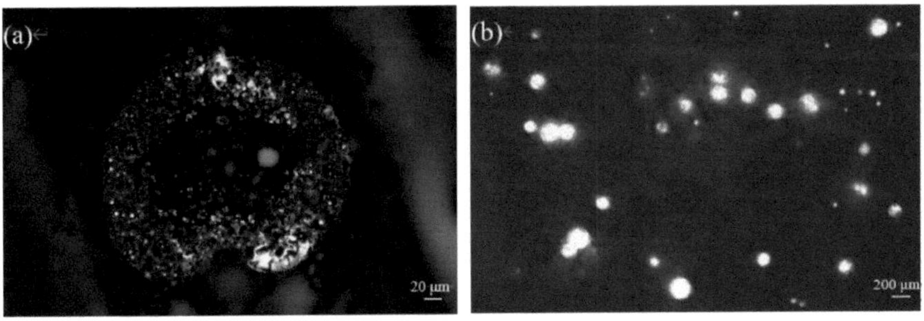

Fig. 4.1 Fluorescence micrographs of the formed OSA by mixing shaking experiments: (a) microplastics; (b) natural sands

of solid–liquid phases (错误!未找到引用源。(a)), which is due to the pore structure and lipophilicity of the microplastics, which adsorb the oil under the action of pore diffusion, liquid film diffusion and surface adsorption, forming the morphology of oil-wrapped microplastics (Yap and Tan 2021). In contrast, the natural sands are wrapped around the exterior of the oil droplet, forming a regular sands-in-oil morphology (错误!未找到引用源。(b)). This is caused by the charge interaction between the particles and the polar part of the crude oil, i.e., the electrostatic force interaction between the negatively charged natural sands and the positively charged asphaltene in seawater (Boglaienko and Tansel 2018; Loh and Yim 2017) (Fig. 1).

4.3.2 *Effect of Suspended Particulate Concentration on Oil Dispersion*

错误!未找到引用源。(a) and (b) plot the variation of *ODE* with time for different initial concentrations of microplastics and natural sands, respectively. When Roncador crude oil was present alone, the *ODE* increased from 13.25% to 31.01% with the mixing time. The *ODE* at stabilization decreased from 6.52% to 3.42% with the increasing microplastic concentration (错误!未找到引用源。(a)), which was much less than the dispersion rate of pure crude oil in water (Yang et al. 2021b). It can also be seen that when the concentration of microplastics was changed, the dispersion effect of crude oil at all concentrations except 25 mg/L did not differ much, with *ODE* ranging from 2.52 to 4.78%; while when the concentration was 25 mg/L, the *ODE* ranged from 5.34 to 6.52%. This indicates that the dispersion effect of crude oil into water does not vary significantly when the microplastics in seawater reach a certain concentration, where the C-H functional groups on the microplastics play a key role in the adsorption of crude oil (Yang et al. 2021a).

The dispersion rates of Roncador crude oil in water were higher than those of pure crude oil at the presence of natural sands (错误!未找到引用源。(b)), which indicates that the natural sands promote the dispersion of oil slicks into water, due to

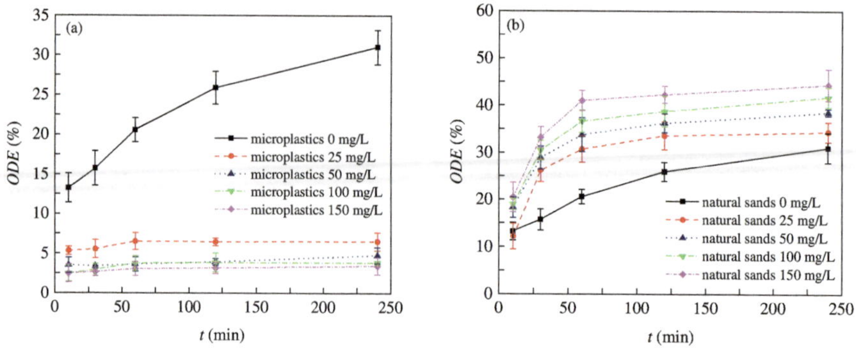

Fig. 4.2 Variation of *ODE* with time under different suspended particle concentrations: **a** microplastics; **b** natural sands

the fact that natural sands have a larger coverage of the surface area of oil droplets, which reduces the structural rigidity by decreasing the oil–water interfacial tension (Yu et al. 2022). Meanwhile, as the concentration of natural sands in the water column increased, the efficiency of crude oil dispersion into water gradually increased and tended to stabilize with the extension of oscillation time. When the dispersion rate reached stable, the *ODE* in water increased from 34.39 to 44.37% with the increase of natural sands concentration (Fig. 2).

4.3.3 Effect of Suspended Particulate Matter Concentration on Oil Particle Size

错误!未找到引用源。(a) and (b) plot the variation of volume median diameter (*VMD*) with time for OSA formed by different particulate matter at different concentrations, respectively. It can be seen that with the increase of oscillation time, the *VMD* of the formed OSA showed a small fluctuation or a smooth decrease until it was stable, while the *VMD* of pure crude oil behaved in the opposite way. The *VMD* of OSA decreased with increasing microplastic concentration (错误!未找到引用源。(a)) (Loh and Yim 2017). At 25 and 50 mg/L, the OSA particle size was larger than that of pure crude oil, ranging from 150 to 180 μm; at 100 mg/L, the OSA particle size did not differ much from that of pure crude oil (Yang et al. 2021b); and at 150 mg/L, the OSA particle size was smaller than that of pure crude oil, at around 135 μm. It is presumed that the larger microplastic concentration makes the microplastics form more buoyant aggregates with oil droplets, while the small particle size OSA is suspended in the water column. The particle size of OSA in water was larger than that of pure crude oil at the start of oscillation, and gradually decreased with the increase of mixing time until the OSA particle size decreased to smaller than that of pure crude oil after 1 h (错误!未找到引用源。(b)). This indicates that at the start of

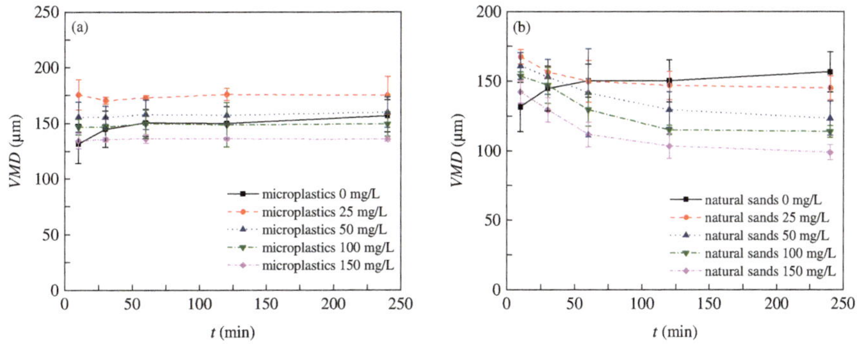

Fig. 4.3 Variation of *VMD* of the formed OSA with time under different suspended sediment concentrations: **a** microplastics; **b** natural sands

the oscillation, the natural sands and crude oil aggregated under the driving of mixing capacity and formed OSA with larger particle size, which was subsequently broken or re-aggregated under the continuous collision of the polymer, and thus suspended in the water or refloated (Fig. 3).

The volume frequency analysis of the OSA particle size that reached the stable form at 240 min of oscillation was performed (错误!未找到引用源。). The percentage of OSA with small particle sizes gradually increased with the increase of microplastic concentration, the peak volume frequency increased from 15.05 to 18.50%, and the corresponding OSA particle size decreased from 217 to 144 μm (错误!未找到引用源。(a)). Meanwhile, the percentage of OSA with large particle size gradually decreased, which indicates that more microplastic fragments aggregated with crude oil, thus forming buoyant OSA and reducing the dispersion efficiency of crude oil in water. The volume frequency of OSA particle size in the presence of natural sands had a small peak at the particle size of 23.4 μm, and the corresponding frequency gradually became larger with the increase of natural sands concentration, from 0.60 to 6.40%; while the opposite pattern is shown at the particle size of 170 μm (错误!未找到引用源。(b)). It is presumed that OSA or oil droplets are more likely to collide and break due to the increase in the number of natural sands in water and the increase in the frequency of mutual collision (Fig. 4).

4.4 Conclusions

When oil spill occurs, the microplastics in the water column can effectively inhibit the dispersion of the oil slick, and gradually form the floating aggregates through the wrapping effect of waves and float on the sea surface, which helps the removal of oil slick. The presence of natural sands accelerates the migration process of oil slick to underwater, and through the breaking effect of waves, the dispersed oil droplets collides and aggregates with natural sands, and the formed aggregates are dispersed

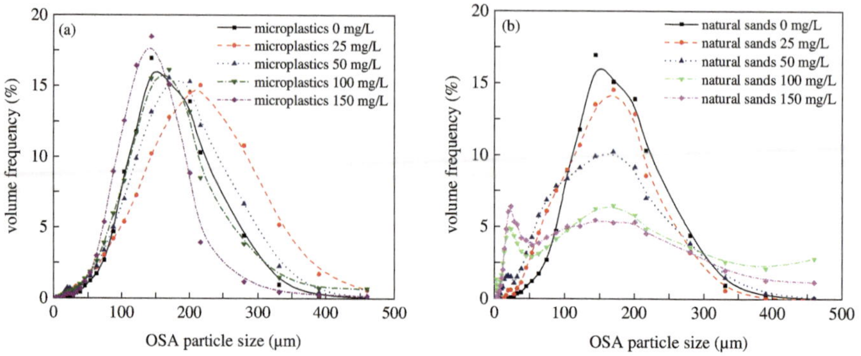

Fig. 4.4 Volumetric frequency distribution of the formed OSA sizes under different suspended sediment concentrations: **a** microplastics; **b** natural sands

into the ocean to reduce the amount of oil slick on the sea surface. By understanding the action mechanism between the microplastics or natural sands and the spilled oil in the water column, it may further improve the capability of responding to marine oil spill accidents, as the aggregation can be potentially served as a technical means of spilled oil removal.

References

Wang F, Gao J, Zhai W et al (2020) The influence of polyethylene microplastics on pesticide residue and degradation in the aquatic environment. J Hazard Mater 394:122517

Thompson RC, Olsen Y, Mitchell RP et al (2004) Lost at sea: where is all the plastic? Science 304(5672):838–838

Moore CJ (2008) Synthetic polymers in the marine environment: a rapidly increasing, long-term threat. Environ Res 108(2):131–139

Browne MA, Crump P, Niven SJ et al (2011) Accumulation of microplastic on shorelines woldwide: sources and sinks. Environ Sci Technol 45(21):9175–9179

Eriksen M, Lebreton LCM, Carson HS et al (2014) Plastic pollution in the world's oceans: more than 5 trillion plastic pieces weighing over 250,000 tons afloat at sea. PLoS ONE 9(12):e111913

Mai L, Bao L, Shi L et al (2018) Polycyclic aromatic hydrocarbons affiliated with microplastics in surface waters of Bohai and Huanghai Seas, China. Environ Poll 241:834–840

Fisner M, Taniguchi S, Moreira F et al (2013) Polycyclic aromatic hydrocarbons (PAHs) in plastic pellets: variability in the concentration and composition at different sediment depths in a sandy beach. Mar Pollut Bull 70(1–2):219–226

Shan J, Wang J, Zhan J, et al (2020) Sorption behaviors of crude oil on polyethylene microplastics in seawater and digestive tract under simulated real-world conditions. In: 25th Topical Meeting of the International-Society-of-Electrochemistry (ISE), pp. 1–8. Chemosphere, Univ Castilla La Mancha, Fabrica Armas, Toledo, Spain

Yang M, Chen B, Xin Y et al (2021a) Interactions between microplastics and oil dispersion in the marine environment. J Hazard Mater 403:123944

Menon NN, Menon NR (1999) Uptake of polycyclic aromatic hydrocarbons from suspended oil borne sediments by the marine bivalve Sunetta scripta. Aquat Toxicol 45(1):63–69

Zhai Y, Xia X, Xiong X et al (2018) Role of fluoranthene and pyrene associated with suspended particles in their bioaccumulation by zebrafish (Dario rerio). Ecotoxicol Environ Saf 157:89–94

Yu Y, Qi Z, Xiong D et al (2021) Oil dispersion and aggregation with suspended particles in a wave tank. J Environ Manage 278:111572

Qi Z, Sun R, Yu Y et al (2021) Performance of dispersed oil and suspended sediment during the oil-sediment aggregation process. Mar Pollut Bull 168:112455

Yap KY, Tan MC (2021) Oil adsorption onto different types of microplastic in synthetic seawater. Environ Technol Innov 24:101994

Boglaienko D, Tansel B (2018) Classification of oil particle interactions in aqueous environments: aggregate types depending on state of oil and particle characteristics. Mar Pollut Bull 133:693–700

Loh A, Yim U (2017) A review of the effects of particle types on oil-suspended particulate matter aggregate formation. Ocean Sci J 51(4):535–548

Yang M, Zhang B, Chen Y et al (2021b) Impact of microplastics on oil dispersion efficiency in the marine environment. Sustainability 13(24):13752

Yu Y, Qi Z, Xiong D et al (2022) Experimental investigations on the vertical distribution and properties of oil-mineral aggregates (OMAs) formed by different clay minerals. J Environ Manage 311:114844

Part II
Meteorology and Bioclimatic Design

Chapter 5
Deviation and Numerical Simulation of Rainfall Forecast in a Warm Sector of Hunan Province

Jinjie Cai, Hongwu Liu, Zhonghai Yin, Wei Fu, and Yan Hu

Abstract From the perspective of duty forecaster, a warm-sector rainstorm process occurred in early summer in Hunan was examined and the method to improve the accuracy was proposed. Based on the conventional observation data, NCEP reanalysis data and FY-2G satellite TBB data, the prediction deviation and its causes are analyzed from the large-scale background and environmental field conditions, and the sensitivity test is carried out using WRF model. The results show that: (1) the process occurred under the circulation background of the subtropical high stretching to the West and lifting to the north, the southwest vortex and the middle and low-level shear moving to the east and pressing to the south, the establishment of the southwest jet and the interaction of the surface cold air. The rainstorm occurred in the warm area on 30 April, with a large prediction bias; On the 1 May, there was cold air, and the southward speed was faster than expected, so the falling area error was also large. (2) There are great differences in the numerical model prediction of heavy rainfall, and the predictability is low. The thermal factor and water vapor factor have little effect on the indication of heavy rainfall, and the dynamic factor is the key. (3) The sensitivity test of WRF model can better predict this process, and it is of reference significance for this type of precipitation forecast in the future.

Keywords Forecast deviation · Warm sector heavy rainfall · Numerical model

---

J. Cai · H. Liu (✉) · Z. Yin · Y. Hu
Hunan Meteorological Observatory, Changsha 410118, China
e-mail: 45877780@qq.com

Hunan Key Laboratory of Meteorological Disaster Prevention and Reduction, Changsha 410118, China

W. Fu
Yongzhou Meteorological Observatory of Hunan Province, Yongzhou 425006, China

C. Yuan et al. (eds.), *Proceedings of 4th International Conference on Resources and Environmental Research—ICRER 2022*, Environmental Science and Engineering,
https://doi.org/10.1007/978-3-031-31808-5_5

5.1 Introduction

The direct or indirect disaster losses caused by heavy rainfall events in warm sector are becoming more and more serious, so it is particularly important to provide accurate and timely heavy rainfall weather forecast services to the society. During the International Labor Day in 2018, there was a heavy rain in Hunan Province, which had a great impact on people's travel. The heavy rainband rapidly pressed southward from the northeast of Hunan. The precipitation error in the warm area on 30 April was large. On 1 May, affected by the cold air, the southward speed was faster than expected, which also led to a large error in the falling area, which had a great impact on people's production, life and travel.

There are many researches on precipitation forecast bias at home and abroad. Zhang et al. (2018) analyzed the forecast departure results of an extreme precipitation in the warm sector in July 2016, and found that the prediction ability of the numerical model for the precipitation process has limit accuracy in this warm sector, which poses a great challenge to the correction ability of the forecaster. CHEN Yun and SUNJisong (Zhan et al. 2012; Sun et al. 2012a, b, 2013; Fang et al. 2012) analyzed the causes of the forecast departure in the 21 July heavy rain process in Beijing from different aspects, and discussed the causes of the extreme precipitation process and the causes of the forecast deviation from the mesoscale environmental conditions, the topographic amplification effect of Taihang Mountain, the precipitation efficiency and other aspects. Fu et al. (2019) made a comparative analysis of the causes of the two cold vortex precipitation in North China and the forecast deviation. It was concluded that the false prediction of precipitation (cold vortex maturity period) mainly came from the numerical model caused by the strong prediction of dynamic conditions, and the missing forecast of precipitation (cold vortex development period) caused by the lack of convective precipitation forecast. Wang et al. (2018) analyzed the deviation of the prediction results of extreme short-term heavy rainfall in Taihang Mountain under the background of weak weather scale, and showed that the main reason for the forecast deviation was that the CAPE value was not referenced, resulting in the underestimate of the instability energy of Shanxi Province, and the thunderstorm downhill enhancement was not predicted. Shang et al. (2017) analyzed the deviation of a rainstorm process in Southern Guizhou caused by the southern branch trough and concluded that the difficulty in the prediction of the rainstorm area lies in the prediction deviation of the establishment, development and impact of the low-level jet. At present, there is little research on the bias of heavy rain event forecast in Hunan, especially the warm-sector heavy rainfall under weak synoptic scale background (Tang et al. 2021). There are also many studies on forecast bias abroad (Ebert and McBride 2000; Hart and Grumm 2001; Junker et al. 2008; Korch 1984; Lalaurette 2003; Wilson et al. 2010). This paper comprehensively analyzes the forecast effect and forecast bias of a heavy precipitation weather process in Hunan from 30 April to 1 May, 2018, and carries out sensitivity tests to find out the reasons for the deviation of precipitation forecast, and try to find a scheme with better forecast method to improve the forecast quality of this kind of precipitation weather.

5.2 Precipitation Situation and Forecast Deviation

From April 30 to May 1, 2018, rainstorm occurred in Hunan Province, and heavy rain weather occurred in some areas (Fig. 5.1a, e). On April 30, heavy rain was mainly located in northern Hunan, with northeast southwest trend (Fig. 5.1a). On May 1, the rain belt pressed rapidly to the south, and the precipitation concentrated in the south of central Hunan, with a quasi east–west trend (Fig. 5.1e). According to the statistics of regional automatic stations from 08:00 on April 30 to 08:00 on May 2 (Beijing time, the same below), the precipitation in 64 towns and townships exceeds 100.0 mm, the maximum 24 h precipitation is 167.6 mm (Huayuan Longtan), and the maximum hourly rainfall intensity is 78.1 mm (01:00–2:00 on the 1st) in Yueshan, Hengpu Township, Linxiang City.

Comparing the rainfall forecast and actual situation of Central China Numerical Prediction Model System (CMA-WRF for short) of the China Meteorological Administration, the Global Assimilation Prediction System (CMA-GFS for short) of the China Meteorological Administration, and the European Center Numerical Prediction (EC for short) 24 h rainfall from 08:00 on the 30th to 08:00 on May 1st which reported since 20:00 on the 29th, it is found that CMA-WRF model predicts no rain or light rain in northern Hunan (Fig. 5.1b), CMA-GFS and EC model predict light to moderate rain (Fig. 5.1c, d), and the rainstorm centers all failed to report. Comparing the rainfall forecast and actual situation from 08:00 on May 1 to 08:00 on May 2 predicted by each central model at 20:00 on the 30th, although CMA-WRF model predicted the rainstorm and heavy rain belt (Fig. 5.1f), the rain belt is northward and eastward, with a deviation of 30 km, which is not referential. CMA-GFS and EC model forecast light rain in central Hunan, with the rainfall falling area slightly north, with a difference of two orders of magnitude (Fig. 5.1g, h), and each model failed to predict the rainfall falling area above the rainstorm in central and

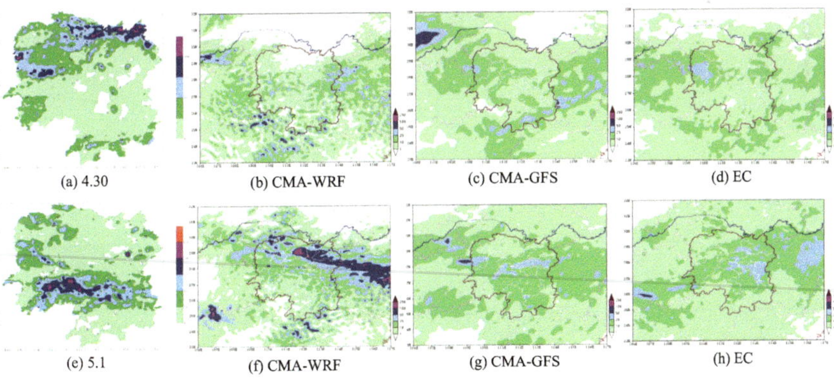

Fig. 5.1 Comparison of cumulative rainfall from 08:00 on 30 April to 08:00 on 1 May 2018, from 08:00 on 1 May to 08:00 on 2 May and 24-h rainfall from predicted by 20:00 on the previous day of each center numerical model forecast system. Unit: mm

southern Hunan. On the whole, all the models have a great deviation from the rainfall intensity and rainfall area of this process.

5.3 Large Scale Circulation Background

Before the torrential rain, the 500 hPa mid-high latitude circulation is a typical 'two troughs and one ridge' circulation pattern in the first rainy season in South China. From the comprehensive map at 20:00 on 30 April, 2018 (Fig. 5.2a), it can be seen that the system configuration of each layer is quite different and the system is relatively shallow. The 5880 geopotential metres of the subtropical high on 500 hPa is located along the coast of South China, the upper trough is located in the northeast of Sichuan, and Hunan is in the southwest airflow in front of the upper trough. At 700 hPa, the shear line is located in the northeast of Hubei to the northeast of Hunan, 850 and 925 hPa shear line is located in the southeast of Chongqing to the northeast of Hubei, Hunan province is located in the south of the shear, controlled by the southwest airflow. The ground is controlled by the warm low pressure (Fig. 5.2c), and the ground convergence line is northeast-southwest to the northeast of Hunan, which provides the trigger condition for the rainstorm in northeast Hunan. On the night of 1 May, the shear line moved eastward and southward with the high-altitude trough. By 08:00 on 2 May, the shear lines at 700, 850 and 925 hPa basically coincided in southern Hunan (Fig. 5.2b), and the surface cold air also began to affect Hunan (Fig. 5.2d, e). Under the combined influence of the upper and lower weather systems, there was a heavy rainstorm in the south of central Hunan at night.

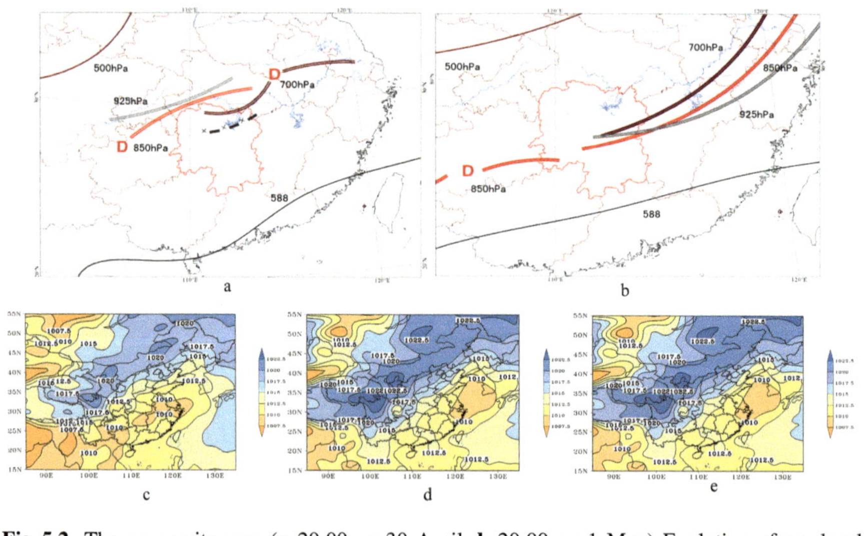

Fig. 5.2 The composite map (**a** 20:00 on 30 April, **b** 20:00 on 1 May) Evolution of sea level pressure (dagpm) (**c** 20:00 on 1 May, **d** 02:00 on 2 May, **e** 08:00 on 2 May)

5.4 Mesoscale Cloud Cluster Characteristics

From the analysis of the evolution diagram of the equivalent blackbody brightness temperature (TBB) of FY-2G, it can be seen that at 18:00 on 30th (Fig. 5.3a), it was the initial stage of the cloud cluster. Convective cloud clusters were formed and developed in southwestern Hubei, southeastern Chongqing and northern Changde, Hunan, and the central brightness temperature was less than-52 °C. At 20:00 on 30th (Fig. 5.3b), the convective cloud cluster in southwestern Hubei weakened in the original place, the convective cloud cluster in southeastern Chongqing moved eastward to Hunan, the cloud cluster in northern Changde also expanded, and rainfall began to occur in northern Hunan. At 01:00 on May 1 (Fig. 5.3c), the cloud cluster in the northwest and northeast of Hunan developed to a strong stage, the intensity and range of MCS reached its peak, and the brightness temperature in the center was less than-58 °C. At this time, the northeast of Hunan was in the gradient large value area of the southwest of MCS. At this time, it corresponded to the strongest period of precipitation in northeast Hunan, and short-term rainstorm occurred. At 03:00, the cloud began to die out, the MCS structure disintegrated, and the intensity gradually weakened. The entire life history of convective clouds lasted about 6 h.

The second day of heavy rainfall occurred mainly in 2:00 to 05:00 on the 2nd May. At 01:00 on February 2 (Fig. 5.4a), the convective cloud cluster began to develop, and the cloud system was formed over Huaihua area; from 02:00 to 03:00 was the development stage of the cloud cluster (Fig. 5.4b), the cloud cluster strengthened continuously and moved slowly eastward and southward; at 04:00 (Fig. 5.4c), the development of cloud cluster was the strongest. There was a clear MCS formation over the north of Shaoyang-Hengyang, which moved slowly, convection developed vigorously, and the center brightness temperature was less than-70 °C. At this time, the middle of Shaoyang was in the gradient large value area of the south of MCS. The short-term rainstorm in Shaoyang and Hengyang was the strongest period of precipitation. From 06:00 (Fig. 5.4d), the central part of MCS began to break, and then the whole cloud belt gradually weakened and disappeared, and the precipitation gradually weakened.

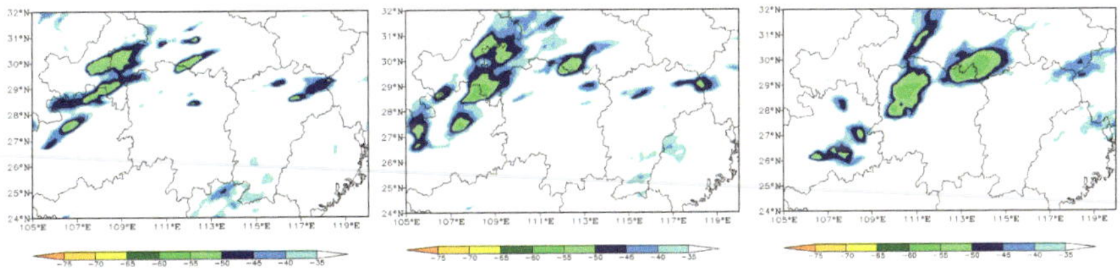

(a) 18:00, 30 April (b) 20:00, 30 April (c) 01:00, 1 May

Fig. 5.3 Evolution of FY-2G equivalent blackbody temperature from 18:00 30 April to 01:00 1 May, 2018 (°C; TBB, shadow)

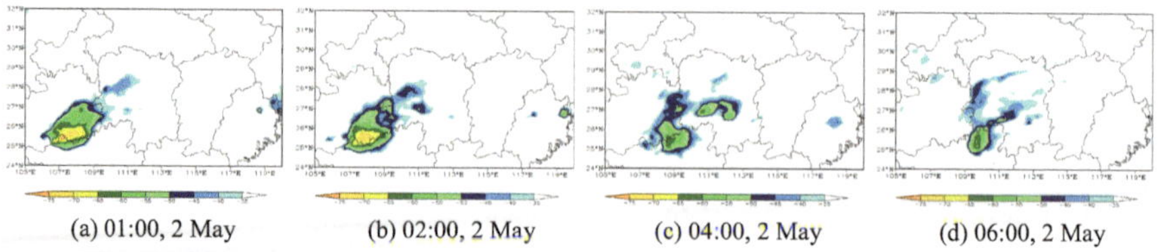

Fig. 5.4 Evolution of FY-2G equivalent blackbody temperature from 01:00 to 06:00 on 2 May, 2018 (°C; TBB, shadow)

5.5 Diagnostic Analysis of Physical Quantity and Rainfall Area

5.5.1 Thermal Factor

The changes of K index with time from 20:00 on 30 April to 20:00 on 1 May were analyzed (Fig. 5.5). At 20:00 on 30 April (Fig. 5.5a), the K index in the north of central Hunan exceeded 35 °C, corresponding to the occurrence of heavy precipitation in this area. At 08:00 on 1 May (Fig. 5.5b), the K index in the south of central Hunan was greater than 35 °C, providing thermal conditions for severe convective weather in the south of central Hunan. The K index decreased rapidly after the rainstorm (Fig. 5.5c).

A vertical cross-section of the temperature advection varying with height along 111°E was made. From 28 to 29 April, most of Hunan Province was controlled by a deep warm-moist air mass (figure omitted). From 30 April, there was a weak

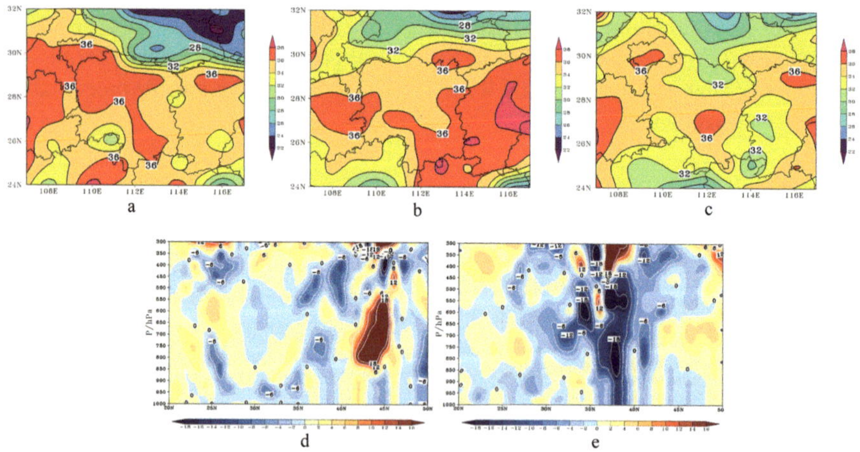

Fig. 5.5 K index (°C) distribution (**a** 20:00 on 30 April, **b** 08:00 on 1 May, **c** 20:00 on 1 May), Distribution of vertical cross section of temperature advection (°C/s) along 111°E (**d** 20:00 on 30 April, **e** 20:00 on 1 May)

cold advection in the upper level (Fig. 5.5d). At 20:00 on 1 May, the range of cold advection in the upper troposphere at 40° N was expanded (Fig. 5.5e), indicating that the cold air moved southward from the upper level into Hunan Province, and the frontal zone was slightly northerly, forming an isothermal advection dense zone between 30° N and 35° N. The warm and humid air climbed up along the isotherm dense zone, and the cold advection column stretched downward (cold advection center value −18 °C), forming a stable precipitation situation with warm front and cold front.

5.5.2 *Water Vapor Condition*

The specific humidity at 850 hPa is shown in Fig. 5.6. During the process, the specific humidity in Hunan was basically maintained at about 12 g/kg. It is characterized by 'high in south and low in north' distribution, and the water vapor flux convergence center is located in the west of Hunan. From 20:00 on 30 April to 08:00 on 1 May (Fig. 5.6a, b), the northeast of Hunan where heavy rainfall occurred was smaller than the humidity, and the convergence of water vapor flux was weak, which had little effect on the warm sector rainstorm. At 20:00 on 1 May (Fig. 5.6c), the specific humidity in southwestern Hunan was close to 14 g/kg, and the low-level water vapor was sufficient, which was also the region with the strongest water vapor flux convergence. This humidity feature was conducive to the occurrence of heavy precipitation in this area. In addition, it can be seen from the characteristics of precipitable water in the whole layer (figure omitted) that the large value area at night of 30 April is concentrated in western and southwestern Hunan, which is not consistent with the large value area of precipitation in northeastern Hunan; the large value area of the whole layer of rainfall on May 1 is located in southern Hunan, which is consistent with the specific humidity conclusion analyzed above.

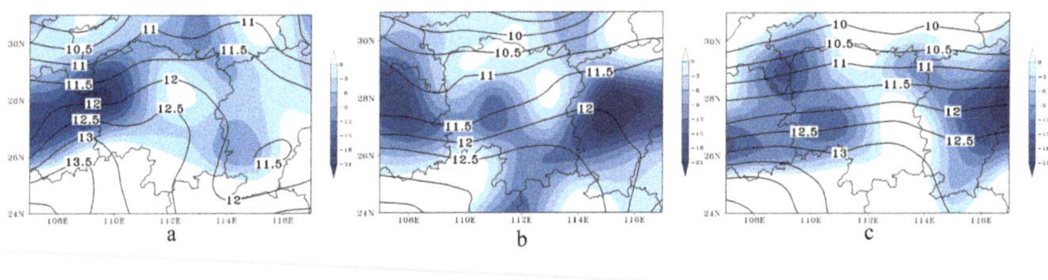

Fig. 5.6 Distribution of 850 hPa specific humidity (g/kg; isoline) and water vapor flux divergence (kg m^{-2} hPa^{-1} s^{-1}; shadow) (**a** 20:00 on 30 April, **b** 08:00 on 1 May, **c** 20:00 on 1 May)

5.5.3 Dynamic Condition (Trigger Mechanism)

The evolution of the surface convergence line at the night of 30 April (Fig. 5.7a–c) was analyzed. The surface convergence line gradually moved from northern Hunan to northeastern Hunan from 20:00, and gradually moved southward from east–west to northeast-southwest, corresponding to the main precipitation area. Subsequently, the surface convergence line continued to press southward, and the area of heavy rainfall was also moving southward. By 20:00 on 1 May (figure omitted), the surface convergence line is located in southwest-northeast direction in southwest Hunan, which has a very good correspondence with the main rainfall area. At the same time, the 850 hPa shear line (Fig. 5.2b) and the surface convergence line are basically superimposed, which is a favorable condition for the maintenance of heavy rainfall, and the surface convergence appears 30 min earlier than the radar echo (Su et al. 2022). Therefore, the location of the convergence line in the surface wind field can be more referenced for the warm sector precipitation. The vertical velocity of 850 hPa was strengthened at the beginning of rainfall (Fig. 5.7d, e), and the rainfall intensity reached the maximum when it developed to the strongest. The upward movement corresponded well with the position of the ground convergence line. Thus, dynamic conditions play a key role in the rainfall process.

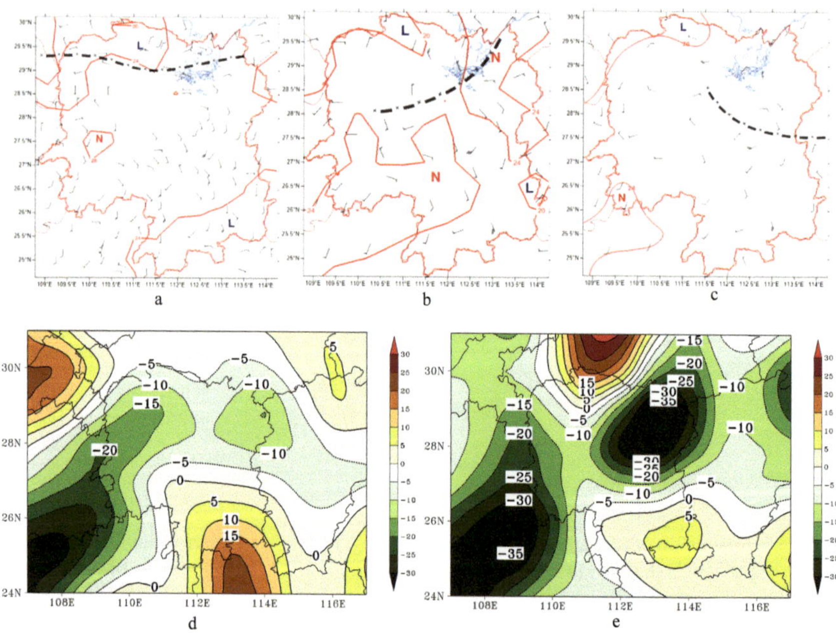

Fig. 5.7 Evolution of surface convergence line in northern Hunan (**a** 20:00 on 30 April, **b** 23:00 on 30 April, **c** 02:00 on 1 May) 850 hPa vertical velocity (10-2 Pa/s) distribution (**d** 20:00 on 30 April, **e** 02:00 on 1 May)

5.6 Numerical Simulation

5.6.1 Model and Data

Using the non-hydrostatic equilibrium mesoscale numerical model WRFv4.0, the initial field and boundary conditions of NCEP (1° × 1° FNL) global analysis data are used. The model area adopts three-layer nested grids. The horizontal grid spacing is 27.9 and 3 km, respectively. The vertical layer is 36 layers, and the model layer top can reach 19 km height. The simulation time is 24 h. The first layer covers most of China, the second layer contains the main precipitation area of this weather process, and the third layer contains the whole province of Hunan.

5.6.2 Precipitation Simulation

Due to the difference of precipitation properties in two days, through the sensitivity experiments of multiple microphysical schemes and cumulus parameterization schemes, compared with the actual situation of Fig. 5.1a and the forecast of other models in Fig. 5.1b–d, it is determined that the precipitation simulation effect of WSM6 microphysical scheme and Kain-Fritsc cumulus parameterization scheme is the best from 08:00 on 30 April to 08:00 on 1 May, 2018 (Fig. 5.8a). Compared with the actual situation of Fig. 5.1e and the forecast of other models in Fig. 5.1f–h, the precipitation simulation effect of WSM6 microphysical scheme and Kain-Fritsch-Cumulus Potential cumulus parameterization scheme is the best from 08:00 on 1 May to 08:00 on 2 May, 2018 (Fig. 5.8b). Therefore, different cumulus parameterization schemes should be adopted for different precipitation processes.

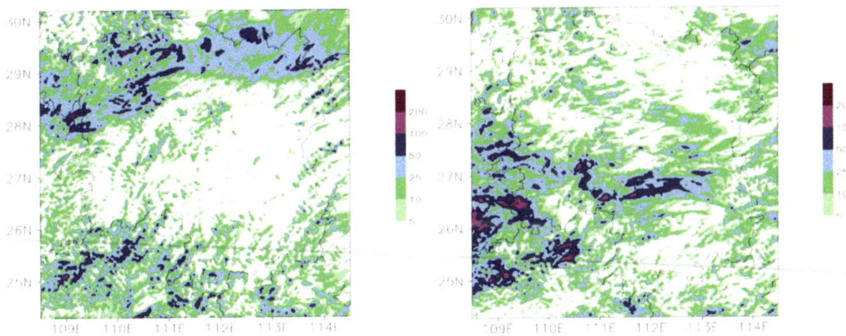

a b

Fig. 5.8 Simulated rainfall (**a** 08:00 on 30 April to 08:00 on 1 May, **b** 08:00 on 1 May to 08:00 on 2 May, unit: mm)

5.7 Conclusions and Discussion

In this paper, the bias and causes of heavy rainfall process from 30 April to 1 May, 2018 are analyzed from the aspects of weather system configuration, physical quantity conditions, mesoscale and smallscale characteristics and model forecast. The following conclusions are obtained:

(1) The heavy rainfall process is the result of the westward and northward movement of subtropical high, the southward movement of mid-low shear, the establishment of southwesterly jet and the surface cold air. The first day (30 April) was a warm sector rainfall, and the forecast bias was large. On the first day, there was cold air influence, but the southward speed was faster than expected, so the forecast bias was also large.
(2) Under the background of weak weather, the indication effect of various physical quantities on heavy rainfall is not great; the predictability of numerical forecast for heavy rainfall in each center is also very low, and the forecast of different models is very different.
(3) In the non-persistent rainstorm process in the flood season, the thermal and water vapor factors have little indication for heavy rainfall, and the dynamic factor is the key.
(4) Different cumulus parameterization schemes can better reflect the different nature of rainfall, which can be used as a reference for forecasters.

Acknowledgements The paper is funded by the following project: China Meteorological Administration Forecaster Special Project (CMAYBY2020-085); Key Fields Research and Development Project of Hunan Provincial Department of Science and Technology (2019SK2161); Key Projects of Hunan Meteorological Bureau (XQKJ22A004); Hunan Business Capacity Building (NLJS14); Hunan Meteorological Bureau Science and Technology Research Project (XQKJ22B005).

References

Ebert EE, McBride JL (2000) Verification of precipitation in weather systems: determination of systematic errors. J Hydrol 239(1–4):179–202
Fang C, Mao DT, Zhang XW et al (2012) Analysis on the mesoscale convections and characteristics of an extreme torrential rain in Beijing on 21 July 2012. Meteor Mon 38(10):1278–1287
Fu JL, Chen S, Shen XL et al (2019) Comparative of the cause of rainfall and its forecast biases of two cold vortex rainfall events in North China. Meteor Mon 45(5):606–620
Hart RE, Grumm RH (2001) Using normalized climatological anomalies to rank synoptic-scale events objectively. Mon Wea Rev 129(9):2426–2442
Junker NW, Grumm RH, Hart R et al (2008) Use of normalized anomaly field to anticipate extreme rainfall in the mountains of Northern California. Wea Forecasting 23(3):336–356
Korch SE (1984) The role of an apparent mesoscale frontogenetic circulation in squall line initiation. Mon Wea Rev 112(10):2090–2111
Lalaurette F (2003) Early detection of abnormal weather conditions using a probabilistic extreme forecast index. Q J R Meteorol Soc 129(594):3037–3057

Shang YY, Du ZJ, Xia XL et al (2017) Forecast deviation analysis: a rainstorm process in Southern Guizhou caused by a small south branch tank. Chin Agric Sci Bull 33(25):127–132

Su AF, Zheng YG, Zhang N et al (2022) A review of research on boundary convergence lines triggering of deep and moist convection. Acta Meteorologica Sinica 80(2):177–189

Sun MS, Li GW, Yin Q et al (2013) Analysis on the cause of a torrential rain occurring in Beijing on 21 July 2012(I): weather characteristics, stratification and water vapor conditions. Torrential Rain Disasters 32(3):210–217

Sun JS, He N, Wang GR et al (2012a) Preliminary analysis on synoptic configuration evolvement and mechanism of a torrential rain occurring in Beijing on 21 July 2012. Torrential Rain Disasters 31(3):218–225

Sun J, Zhan Y, Yang SN et al (2012b) Analysis and thinking on the extremes of the 21 July 2012b torrential rain in Beijing. Part (II): preliminary causation analysis and thinking. Meteor Mon 38(10):1267–1277

Tang J, Ye CZ, Tang MH et al (2021) Contrast analysis of two warm-sector heavy rainfall processes under weak synoptic scale background in Hunan. Plateau Meteorol 40(4):815–828

Wang CM, Yu XD, Liu J et al (2018) Analysis of a forecast failure case of extreme flash-rain under weak synoptic-scale background in Taihang Mountain. Meteor Mon 44(1):107–117

Wilson JW, Feng YR, Chen M et al (2010) Nowcasting challenges during the Beijing Olympics: successes failure, and implications for future nowcasting systems. Wea Forecasting 25(6):1691–1714

Zhan Y, Sun J, Xu J et al (2012) Analysis and thinking on the extreme of the 21 July 2012 torrential rain in Beijing. Part I: observation and thinking. Meteor Mon 38(10):1255–1266

Zhang PP, Sun J, Dong LP et al (2018) Analysis of extreme precipitation forecast deviation in a warm region of Hubei province in 2016. J Meteorol Environ 34(5):01–08

Chapter 6
The Analysis of Forecast Bias and Causes of Extreme 'Dragon Boat Water' in Hunan Province in 2022

Yan Hu, Enrong Zhao, Hongwu Liu, Hui Zhou, and Jinjie Cai

Abstract An extreme rainstorm weather process occurred in Hunan during the Dragon Boat Festival from June 1 to June 5, 2022, which was the strongest 'Dragon Boat Water' in the past 20 years, causing serious secondary disasters. Based on the conventional observation data, NCEP reanalysis data, FY-4 satellite TBB data and HYSPLITv4.9 trajectory tracking model, the prediction deviation and causes of the rainstorm are analyzed from the large-scale background, mesoscale characteristics and water vapor transport. The results show that: (1) Extreme 'Dragon Boat Water' can be divided into two distinct rainfall stages: warm-sector torrential rain and frontal torrential rain. In the first stage, the shear line weakened at daytime and strengthened at night increased the difficulty of forecast. In the second stage, the cold air and shear made several β, γ scale clouds develop and merge into medium α scale clouds, resulting in systematic heavy rainfall. (2) The intensity of water vapor convergence in the first stage was greater than that in the second stage, which was one of the factors that the maximum hourly rainfall intensity appears in the first stage. (3) There were four water vapor transport channels of the extreme rainstorm, the main was the Arabian Sea. There was a rare short-distance cold and humid water vapor transport channel near the Hetao area (accounting for 20%). Forecasters seldom pay attention to the northern water vapor transport, which may be one of the important reasons for the deviation of the extreme rainstorm forecast.

Keywords Extreme 'Dragon Boat Water' Process · Forecast deviation · Mesoscale characteristics · Water vapor characteristics

---

Y. Hu · E. Zhao (✉) · H. Liu · H. Zhou · J. Cai
Hunan Meteorological Observatory, Changsha 410006, China
e-mail: 410083029@qq.com

Hunan Provincial Key Laboratory of Meteorological Disaster Prevention and Mitigation, Changsha 410006, China

C. Yuan et al. (eds.), *Proceedings of 4th International Conference on Resources and Environmental Research—ICRER 2022*, Environmental Science and Engineering,
https://doi.org/10.1007/978-3-031-31808-5_6

6.1 Introduction

Extreme rainstorm is difficult to forecast because of its characteristics of strong paroxysm, long duration of heavy rainfall, overlap of rainfall areas and large hourly rainfall intensity, which can easily lead to serious disasters such as urban waterlogging, basin flood and flash flood, it poses a great threat to the safety of people's lives and property. Experts have carried out a lot of research work on rainstorm forecasting technology, and studied the causes of extreme rainstorm from the aspects of circulation anomaly, multi-scale characteristics and water vapor transport. Liu et al. (1999). found that the frequency of heavy rainfall in the Yangtze River basin was increasing year by year, and the rainfall intensity was the highest in summer. Bao and Huang (2006). concluded that the middle and lower reaches of the Yangtze River is the area with the most summer heavy rainfall in China by analyzing the climatic characteristics of 40a rainfall data. Zhang et al. (2007). used wavelet analysis method to study the evolution characteristics of extreme precipitation in flood season in the Yangtze River Basin. It was found that the extreme precipitation in most areas was increasing, and the maximum value of extreme precipitation in Hunan appeared in June. Water vapor transport plays an important role in the evolution of precipitation process, and the difference of transport path has different contribution to the rainstorm area (Trenberth 1998; Jiang et al. 2011; Hu et al. 2022). During the '7.21' extreme precipitation event in Henan in 2021, the hourly rainfall intensity in Zhengzhou reached 201.9 mm, which broke through the hourly rainfall extreme value of the land observation station (Su et al. 2021). In consideration of the fact that the Lagrangian method can provide details and changes along the water vapor transport trajectory, some scholars have applied this method to study the water vapor source and budget of rainstorms in recent years (Shi et al. 2022; Stohl and James 2005; Dirmeyer et al. 2009). Yu et al. (2022). simulated the trajectory of the air mass in the Zhengzhou rainstorm center through the HYSPLIT model and concluded that the water vapor mainly came from the South China Sea and the North Pacific Ocean. Duan et al. (2022). found that the southeast jet from the edge of the subtropical high is an important water vapor transport channel in the central region of Henan. Zhou et al. (2022). Analyzed several extreme rainstorms in the middle and lower reaches of the Yangtze River and found that the continuous influence of the mesoscale convective system (MCS) in the boundary layer was an important cause of the heavy rainfall. When Huang et al. (2021). discussed the extreme heavy rainfall process in the middle reaches of the Yangtze River, it was concluded that the multiple convective clouds on the Yangtze-Huaihe shear line affected the eastern Hubei region with the train effect, which was the cause of the persistent rainstorm.

According to the statistical characteristics of extreme heavy rainfall in Hunan, the number of extreme heavy rainfall days and rainfall increased significantly since 1993 (Luo et al. 2008), and the daily extreme rainstorm gradient has increased. Northeast and northwest Hunan are two extreme rainstorm centers in Hunan (Zhang et al. 2012). An extreme heavy rainfall occurred in Hunan from June 1 to 5, 2022. It has the characteristics of long duration, large cumulative rainfall, strong extremeness and high overlap of falling area. It is an extremely strong regional rainstorm, causing basin flood, debris flow, urban waterlogging and other disasters. Heavy rainfall period at the Dragon Boat Festival holiday, regional average daily rainfall and the Dragon Boat Festival 5-day average rainfall are ranked fourth in history, for the extreme 'dragon boat water'. This paper analyzes the causes from the interaction of high and low altitude systems, water vapor transport characteristics, and dynamic characteristics, in order to find out the reason of precipitation forecast deviation and provide some reference for future extreme rainstorm forecasts.

6.2 Rainfall and Disaster

June 1, 2022 08:00 to 6 08:00, the average rainfall in Hunan Province was 83.6 mm (Fig. 6.1). The cumulative rainfall of 1046 stations in 89 counties and cities exceeded 100 mm, covering an area of 72,000 km^2 (accounting for 33.94% of the province's area), and the maximum cumulative rainfall was 400.5 mm (Huayuan Yayou Town Station). From 08:00 on June 2 to 08:00 on June 3, the rainstorm and heavy rainstorm stations were the most, with 163 stations of heavy rainstorm, 312.5 mm of maximum daily rainfall (Taojiang Santangjie Town Station), 94.9 mm of maximum hourly rainfall intensity (Mayang Longjiabao Township, from 01:00 on June 2 to 02:00 on June 3). A total of 7 counties rainfall extreme precipitation standards.

The continuous heavy rainfall led to a basin-wide flood process in the middle and lower reaches of the Yuanshui River, Zishui River, Xiangjiang River and Dongting Lake area. The first flood occurred in Yuanjiang River and Zishui River in 2022, with 10 stations exceeding the warning water level and 2 stations exceeding the guaranteed water level. The maximum inflow of the Wuqiangxi reservoir in Yuan River was 25,400 cubic metres per second at 2:00 on June 3. The peak water level at the Zishui Taojiang County hydrologic station was 39.56 m at 8:30 h on June 3, exceeding the warning level of 0.36 m. There were 1,590,200 people affected, 103,000 hectares of crops were affected, 605 houses of 210 families collapsed, 1,759 houses of 702

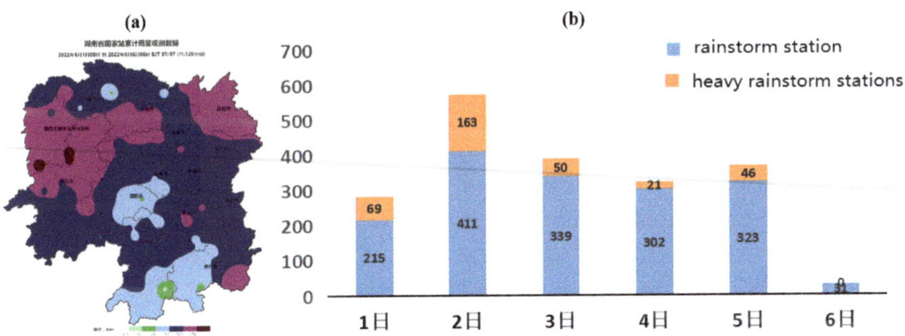

Fig. 6.1 Accumulated rainfall (**a**, unit: mm) and time series of rainstorm daily station (**b**, unit: number) from 08:00 on 1 to 08:00 on 6 June 2022 in Hunan. Unit: mm

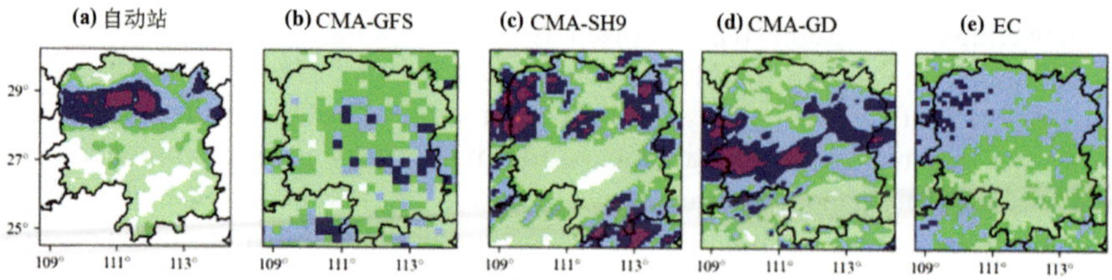

Fig. 6.2 Accumulated rainfall (**a**) and accumulated precipitation by models (initial: at 20:00 BT 1 June 2022, **b**: CMA-GFS; **c**: CMA-SH9; **d**: CMA-GD; **e**: ECMWF) from 08:00 on 1 to 08:00 on 6 June 2022 in Hunan. Unit: mm

families were seriously damaged, and 110 landslides, mudslides and other geological disasters were caused, the direct economic loss was 2.794 billion yuan.

On June 2, the strongest rainfall period of this process, the subjective and objective forecast were deviated. The 24-h forecast rainfall of each model from 08:00 on June 2 to 08:00 on June 3 is compared with the actual situation (Fig. 6.2). The rainstorm and heavy rainstorm appear in the central and western Hunan. The global model CMA-GFS and ECMWF forecast areas are more consistent, and the rainfall level is smaller. The regional model CMA-SH9 can better predict the heavy rainstorm center, but the range is larger. The CMA-GD forecast precipitation center is located southward.

6.3 Data and Method

This paper selects three types of data: real-time monitoring data of hourly rainfall at regional stations of China Meteorological Administration in June 2022; NCEP reanalysis data of 6 h step from National Centers for Environmental Prediction; GDAS data for trajectory tracking model provided by National Centers for Environmental Prediction (converted to Beijing time); national Satellite Meteorological Center FY-4 satellite hourly TBB data.

HYSPLIT v4.9, an air mass trajectory tracking model based on Lagrange method, is used to tack and simulate the source and evolution of water vapor in the rainstorm area. This trajectory model is developed by NOAA and the Australian Bureau of Meteorology (Draxler and Hess 1998). By running all trajectories, the total spatial variance (TSV) is calculated according to cluster pairing and multiple iterations, and the number of clustered trajectories is determined by the TSV variation. The clustering method can well obtain the contribution rate of different water vapor transport channels to the rainstorm area (Sun and Wang 2014, 2015). The map involved in this paper is based on the map review numbers GS (2016) 2948 and GS (2016) 1550 standard maps downloaded from the standard map service website of the National Bureau of Surveying, Mapping and Geographic Information, and the base map is unmodified.

6.4 Cause Analysis of Deviation

6.4.1 Circulation Features

The analysis of 500 hPa average circulation and anomaly field (Fig. 3a) shows that there is a meridional circulation of 'two troughs and one ridge' in the middle and high latitudes of the middle and upper troposphere. The eastern part of Baykal maintains a strong blocking high. The average positive anomaly in the blocking high is 10–20 dagpm, and the northeast cold vortex develops deeply. At low latitudes, the western Pacific subtropical high controls most of the southern part of South China in an east–west block distribution, and its intensity is slightly weaker than that in the same period of normal years, which is conducive to the eastward outflow of the plateau shortwave trough. The development of the process can be divided into two stages. The first stage is from June 1 to 3 (Fig. 3b, c). In the middle and high latitudes, two troughs and one ridge are maintained stably, and the main body of the subtropical high retreats slightly eastward, and the western ridge point of 588 dagpm retreats from 120 °E to 126 °E. In the positive vorticity advection area in front of the shortwave trough in Hunan and the warm and humid unstable airflow at the northern edge of the subtropical high 584 dagpm, there are southwest jets at 700 hPa and 850 hPa in the middle and lower layers, and the jet wind speed reaches 16–20 m/s. At 850 hPa, there is a stable southwest vortex at the junction of Hunan and Guizhou. The cyclonic vortex extends to the middle and upper troposphere, and the vortex is deep. The position of the shear line in the east section is less moving, extending to the west and north of Hunan, resulting in regional continuous rainstorm. The second stage is from June 4 to 5 (Fig. 3d, e). At 08:00 on June 4, the subtropical high moved westward and strengthened, and the plateau trough moved eastward rapidly, which drove the southwest vortex to move into Hunan. The southwest jet on the southeast side of the vortex center strengthened to 20 m/s, and the surface cold air invaded, the cold and warm met, and systematic precipitation occurred in the north of central Hunan. By the day of June 5, with the eastward movement of the southwest vortex and the southward movement of the cold front, the heavy rain belt moved rapidly to southern Hunan.

It can be seen that the precipitation in the first stage of the process is a warm-sector rainstorm, the main influence system is the shear line in the eastern section of the vortex, the edge of the subtropical high and the southwest jet in the middle and lower troposphere increase the instability, and the numerical model forecast has bias.

6.4.2 Characteristics of Mesoscale Convection

In meteorology, a cloud cluster with a cloud top brightness temperature TBB (Black Body Temperature) ≤ -52 °C, an area greater than 5000 km^2, an eccentricity greater

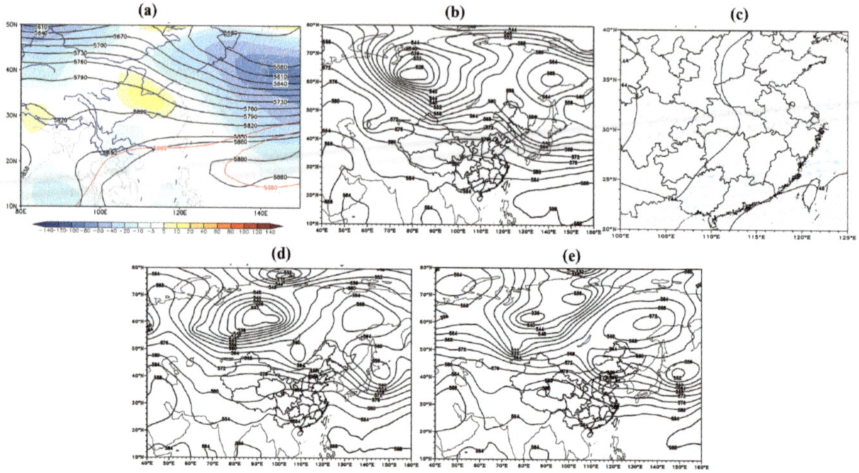

Fig. 6.3 The average height field at 500 hPa (black contour, unit: dagpm) and anomaly field (color area, unit: dagpm) from 1 to 6 June 2022 (**a**), the geopotential height field on 500 hPa (**b** 20:00 on 2 June; **d** 08:00 on 4 June; **e** 08:00 on 5 June, unit: dagpm) and the average wind field at 850 hPa (**c** 20:00 on 2 June, unit: ms^{-1})

than or equal to 0.7, and a duration greater than 6 h is defined as a mesoscale convective complex (MCC) (Maddox 1980). The hourly TBB evolution was analyzed for the two stages of the 'Dragon Boat Water' process (Fig. 6.4). In the first stage, the vortex center was located in the central part of Guizhou, the wind shear in western Hunan increased from 2 to 12 m/s due to the strengthening of the shear in the eastern section of the vortex at night on June 1, and the convective clouds of medium γ and medium β scales were continuously generated. At 23:00, the cold cloud (TBB ≤ −52 °C) is expanding constantly and developing into an isolated medium α scale cloud cluster, with a width of about 170 km and a length of about 320 km, and the minimum TBB is −60 °C. From 23:00 on June 1 to 02:00 on June 2, 29 stations in Xiangxi and Huaihua appeared heavy rain, and 3 stations appeared extremely heavy rainstorm, the maximum precipitation in 3 h is 183.9 mm (Mayang Longjiapu), and the cold cloud with TBB ≤ −60 °C lasted for 9 h, resulting in the strongest convective rainfall period. On June 2, it weakened rapidly in the daytime. At night, as the vortex shear line extended eastward, the mid-troposphere southwesterly flow strengthened again, and the shear from northern Guizhou to northern Jiangxi increased to over 10 m/s. The cold cloud cluster in the central and eastern of Hunan is forming and developing, and the area of TBB ≤ −52 °C cold cloud cluster is expanding rapidly. At 08:00 June 3, several medium α scale MCC clouds developed and merged from central Guizhou to central Hunan, forming distinct east–west banded shear line clouds. The lowest TBB of cold cloud is −80 °C and maintained for more than 8 h, the ≤−52 °C cold cloud cluster lasted for more than 22 h. The heavy rainfall area was stable in western Hunan, causing 16 stations 3 h rainfall more than 100 mm, rainstorm center was mainly located in the northern foot of Xuefeng Mountain. The shear weakened

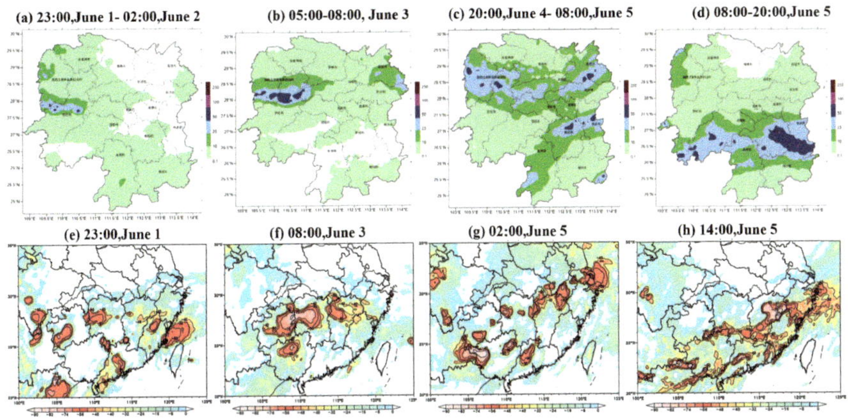

Fig. 6.4 The accumulated rainfall (the first stage: **a~b**, the second stage: **c~d**) (color area, unit: mm) and the black body temperature (TBB) at different time (the first stage: **e~f**, the second stage: **g~h**) (the shaded, unit: °C) in Hunan from 1 to 5 June 2022

and the rainfall significantly weakened from early morning to daytime on June 4. It can be seen that the shear cloud system developed strongly and influenced for a long time, and the development of MCC from isolated to organized is also one of the important causes of this extreme rainstorm.

In the second stage from 20:00 on June 4 to 08:00 on June 5 (Fig. 4g, h), The shear line strengthened again, the eastern section moved northward to the middle and lower reaches of the Yangtze River, and the western segment is maintained in central Hunan, where several scattered convective clouds develop along the shear line, with TBB between −40 and −50 °C. The precipitation intensity slightly weakened, mainly stable rainfall, 12-h rainfall 178 stations greater than 50 mm. on June 5 daytime, the shear line moved southward to the south of Hunan, the surface cold air moved southward rapidly, and the cold air climbed on the southwest airflow. The cold and warm air converged, the convergence ascending motion strengthened, and the heavy rainfall center moved southward rapidly. The minimum TBB reached −60 °C at 14:00 in southern Hunan, and there were 207 stations rainfall more than 50 mm and 39 stations greater than 100 mm from 08:00 to 20:00 on June 5. The center of the rainstorm was mainly located in the northern foot of the Nanling Mountains. It can be seen that the combined effect of the cold air and shear is the main reason for the second stage of systematic heavy rainfall.

6.4.3 Vapor Condition

Moisture Transport Path. The vertical integral regional average of water vapor flux and water vapor flux divergence from the ground to 300 hPa in the two stages of the process are selected respectively (Fig. 6.5). The analysis shows that the main water vapor transport belt in the first stage of this process is a warm and wet from the Arabian Sea, and around the southern part of the Indian Peninsula. The water vapor increases in the Bay of Bengal and turns to the southwest. There is a slight loss of water vapor when it crossed the Indo-China Peninsula. It meets with another southerly airflow from the South China Sea near Hainan Island. The two water vapor channels are superimposed and transmitted to the western region of Hunan Province in a southwest path. The intensity of the whole layer of water vapor flux is strengthened. The maximum water vapor flux in the rainstorm area reaches 800 $kg \cdot m^{-1} \cdot s^{-1}$, and the maximum water vapor flux divergence near the shear line in the eastern section of the vortex is -16×10^{-6} $g \cdot m \cdot kg^{-1} \cdot s^{-2}$. Strong water vapor convergence. In the second stage, the water vapor mainly comes from a strong southwest airflow from the Bay of Bengal, which directly flows into the rainstorm area after passing through the Indo-China Peninsula and the South China Sea. There is also a northwest airflow from northern Xinjiang, the water vapor transport is weak. It travels southward from the Hetao area in the northwest path and transports to eastern Hubei, and then transports to Hunan in the northerly airflow. This airflow transport intensity is weaker than the southwest water vapor channel, and the two airflows intersect in southern Hunan. The water vapor convergence zone is narrowly distributed from the middle and lower reaches of the Yangtze River to southern Hunan in a northeast-southwest direction, and the maximum water vapor convergence center is -10×10^{-6} $g \cdot m \cdot kg^{-1} \cdot s^{-2}$.

Water Vapor Contribution Rate. From the time series change of the average water vapor flux divergence profile in the rainstorm area of this process (figure omitted), it is found that the water vapor in the middle and lower troposphere converges strongly, and the time period of the water vapor convergence center is consistent with the heavy rainfall period. The maximum water vapor convergence level is located at 800 hPa, and the water vapor accumulates in the lower layer, which is closely related to the strengthening of the southwest airflow on the south side of the shear line. Using HSYPLIT4.0 model, according to the above method,

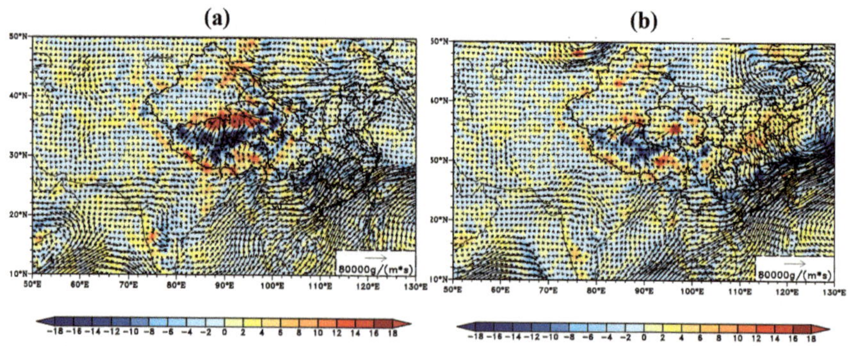

Fig. 6.5 The whole layer moisture flux moisture flux divergence in 'Dragon-boat Rain' process (**a** the first stage: from 08:00 on 1 to 08:00 on 4 June; **b** the second stage: from 20:00 on 4 to 20:00 on 5 June 2022) Black arrows denote the moisture flux (**a** unit: $\times$ 105 $g \cdot (cm \cdot s)^{-1}$), Color shaded represent the moisture flux divergence (unit: $\times$ 10^{-6} $g \cdot m \cdot kg^{-1} \cdot s^{-2}$)

the rainstorm center is selected as the simulation area (27.5° N–30° N, 108° E–114° E). The simulation started at 08:00 on June 6, with 800 hPa as the initial height of the simulation. Considering the long duration of rainfall, the model gives hourly trajectory points backward to 10 days by re-tracking every hour. All trajectories were clustered. According to the spatial variance growth rate (TSV) (Fig. 6a), there is a sharp increase when the four trajectories are clustered, and the contribution rate of the four water vapor channels to the rainstorm area is determined and analyzed.

From the results of 800 hPa air cluster and the change of water vapor channel with height (Fig. 6c, d), there are four main water vapor transport channels of the extreme 'Dragon Boat Water' rainstorm, the most important channel is from the Arabian sea (channel 1), the water vapor contribution rate is 36.9%, channel 1 come from the ocean surface in low latitude with trajectory height around 1000 hPa, at first it transport along the Bay of Bengal sea surface in southwest path, then through the Indo-China Peninsula and the Yunnan-Guizhou Plateau, the air mass continues to climb and finally transports water vapor to the rainstorm area near 800 hPa. The second is from the Bay of Bengal (channel 2), the water vapor contribution rate is 32.14%, channel 2 also comes from the ocean surface in low latitude with trajectory height around 1000 hPa, at first it transports to the south sea in southwest path, then Convert to the southerly path through Guangxi, Guizhou, air mass climb with the terrain height in the land, climbed through the Xuefeng Mountains in western Hunan

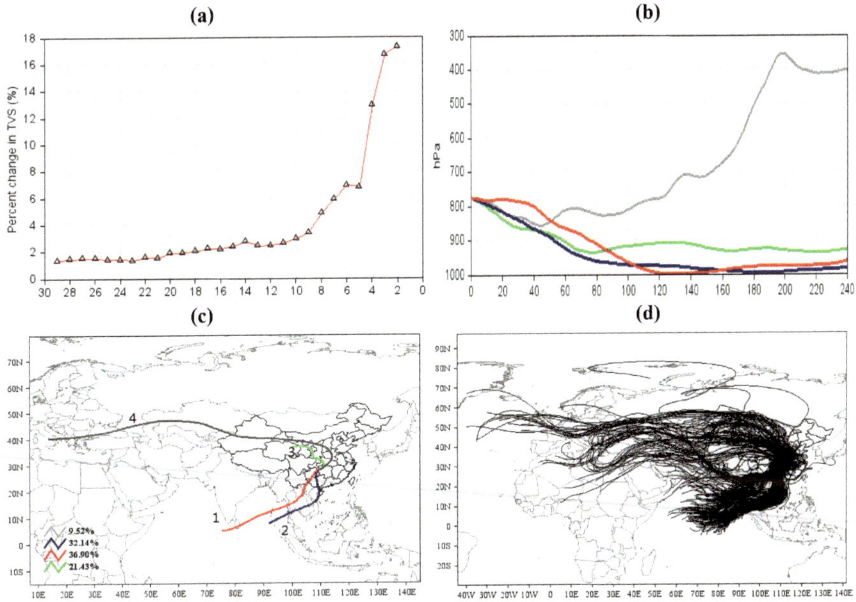

Fig. 6.6 The increase rate in the total spatial variance (TSV) as cluster combined at 800 hPa from 1 to 6 June (**a**), the time-height section of the air parcels the trajectory clustering at 800 hPa (**b**), the corresponding percentage of the number of tracks at 800 hPa (**c**), back trajectories of moisture of all water vapor transmission path (**e**) at the Hunan region from 1 to 6 June 2022

when the height of the fastest rising. The third one is a short water vapor transport channel from the west of Hetao area (channel 3), the water vapor contribution rate is 21.43%, the air mass originated from the land, with an initial height of about 800 m. Blocked by the Qinling Mountains transmission speed is slow, the height of the air mass gradually increased and was transported to the rainstorm area in a northerly path. This short-range water vapor transport channel is rare for the rainstorm process in Hunan, which may be affected by the eastward outflow of the north branch trough and the rich water vapor brought by the southwest airflow in front of the trough. The channel 4 from the southern Europe is longest, with an initial height of about 7000 m. The air mass pass through eastern Europe, western Asia and enter into Xinjiang. Then the air mass transport to the North China Plain by the east path, the height decreases rapidly, and then transport to the rainstorm area by the northeast path. The water vapor contribution rate is 9.52%, it is mainly the dry cold air mass from the north, which provide little water vapor. From the above, we know that the main water vapor source during the extreme 'dragon boat water' process in Hunan are the Arabian Sea and the Bay of Bengal, indicating that due to the weak and southerly main body of the subtropical high, the southwest warm and humid airflow in front of the trough provides a large amount of water vapor for the torrential rain area.

6.5 Conclusion

Extreme 'Dragon Boat Water' rainstorm process in Hunan in 2022 is the strongest precipitation process during the dragon boat festival in recent 20 years. There is a certain deviation in the forecast. This paper analyzes the deviation and discusses the causes from the aspects of circulation background, mesoscale characteristics and water vapor conditions. The results show that:

(1) the extreme 'Dragon Boat Water' can be divided into two distinct rainfall stages, warm-sector torrential rain and frontal torrential rain. In the first stage, the shear line weakened at daytime and strengthened at nighttime increased the difficulty of forecasting. In the second stage, the low trough drove the shear line move eastward and southward, the cold air moved southward, the cold and warm met, the strong rain belt showed significant mobility, and the forecast deviation was small.

(2) In the first stage, the shear cloud system developed strongly and lasted for a long time. The minimum value of TBB appears a low center of −80 °C. The organized development of MCC is also an important cause of insufficient consideration of the extreme rainstorm magnitude. In the second stage, the combined effect of cold air and shear make several β, γ scale clouds develop and merge into medium α scale clouds, resulting in systematic heavy rainfall.

(3) The water vapor convergence intensity in the first stage is greater than that in the second stage. There are differences in water vapor source and transport path between the two stages, especially in the second stage, there is also a water vapor transport channel from the north.
(4) There are four main water vapor transport channels of the extreme 'Dragon Boat Water' rainstorm, The most important channel is from the Arabian sea, the second is from the Bay of Bengal, the water vapor contribution rate of the above two water vapor channels is 68%. There is also a rare short-distance water vapor transport channel near the Hetao area, the water vapor flux contribution rate of this cold and wet channel is also more than 20%. Forecasters rarely pay attention to the water vapor transport in the north, which may also be one of the important reasons for the deviation of this extreme rainstorm forecast.

Acknowledgements Key Fields Research and Development Project of Hunan Provincial Department of Science and Technology(2019SK2161); Key Projects of Hunan Meteorological Bureau (XQKJ22A004); China Meteorological Administration Innovation and Development Project (CXFZ2021Z033, CXFZ2021J020); Hunan Business Capacity Building (NLJS14); Hunan Meteorological Bureau Science and Technology Research Project (XQKJ22B005); Hunan Meteorological Bureau Science and Technology Research Project (XQKJ21C010).

References

Bao M, Huang RH (2006) Characteristics of the interdecadal variations of heavy rain over China in the last 40 years. Chin J Atmos Sci 30(6):1057–1067
Dirmeyer PA, Schlosser CA, Brubaker KL (2009) Precipitation, recycling, and land memory: an integrated analysis. J Hydrometeorol 10(1):278–288
Draxler RR, Hess GD (1998) An overview of the HYSPLIT_4 modeling system for trajectories, dispersion and deposition. Aust Meteorol Mag 47:295–308
Duan T, Chen QL, Liao YJ (2022) Analysis of "21.7" extreme rainstorm formation process and disaster mechanism in Zhengzhou. J Meteorol Sci 42(2):152–161
Hu Y, Liu HW, Zeng YT et al (2022) Analysis of water vapor transport for disaster of two different types of rainstorms in Hunan. J Catastrophol 37(3):93–98
Huang ZY, Wang JY, Zhou W (2021) Characteristics analysis of an extreme heavy rain event in the middle reaches of the Yangtze River from July 4 to 8 in 2020. Torrential Rain Disasters 40(4):333–341
Jiang ZH, Liang ZR, Liu ZY et al (2011) A diagnostic study of water vapor transport and budget during heavy precipitation over the HuaiHe River basin in 2007. Chin J Atmos Sci (in Chinese) 35(2):361–372
Liu XN (1999) Climatic characteristics of extreme rainstorm events in China. J Catastrophol 14(1):54–59
Luo BL, Zhang C, Lin H (2008) Characteristics of climatic changes and catastrophe of extreme precipitation in Hunan in recent 40 years. Meteorol Monthly 34(1):80–85
Shi Y, Jiang ZH, Li ZX (2022) Vertical characteristics of water vapor transport during the rainy season in eastern China based on the Lagrangian method. Chin J Atmos Sci 46(2):380–392
Stohl A, James P (2005) A Lagrangian analysis of the atmospheric branch of the global water cycle. Part II: Moisture transports between earth's ocean basins and river catchments. J Hydrometeorol 6(6):961–984

Su AF, Lv XN, Cui LM et al (2021) Predicition and test of optimal integrated precipitation based on similar spatial distribution of precipitation. Torrential Rain Disasters 40(5):445–454

Sun B, Wang HJ (2014) Moisture sources of semiarid grassland in China using the Lagrangian particle model FLEXPART. J Climate 27(6):2457–2474

Sun B, Wang HJ (2015) Analysis of the major atmospheric moisture sources affecting three sub-regions of East China. Int J Climatol 35(9):2243–2257

Trenberth KE (1998) Atmospheric moisture residence times and cycling: implications for rainfall rates and climate change. Clim Change 39(4):667–679

Yu TF, Li C, Shi J (2022) Analysis on water vapor transport characteristics of "7.20" extreme rainstorm in Zhengzhou. J Mar Meteorol 42(3):69–76

Zhang W, Shou SW, Yang JH (2007) Analysis of anomaly characteristics of extreme precipitation in flood season in the mid-lower reaches of Yangtze River. Meteorol Monthly 33(3):61–67

Zhang JM, Liao YF, Duan LJ et al (2012) Climate variations of extreme continuous rainstorms in Hunan during 1960–2009. Geogr Res 31(6):1004–1015

Zhou JL, Zhang JG, Wu T et al (2022) Characteristics of the mesoscale weather system producing extreme rainstorm in boundary layer during the Meiyu front over the middle reaches of Yangtze River. Meteor Mon 48(8):1007–1019

Chapter 7
A Non-geostrophic Wet Q-vector Analysis of Continuous Heavyrain in Hunan

Haichao Cai, Jinjie Cai, Zhonghai Yin, Hongwu Liu, and Lu Wang

Abstract Using NCEP $1^\circ \times 1^\circ$ reanalysis data, Conventional meteorological observation data and surface intensive observation data, the non-geostrophic Q vector theory, diagnostic analysis is performed in a continuous rainstorm process in Hunan from 2012 June 25 to 28. The results show that: the change of non-geostrophic Q-vector convergence zone is a good indication of the heavy precipitation area, stable and less strong Q-vector convergence center for persistent heavy rainfall occurred; Rainstorm peak, with a plurality of Q-vector divergence near the middle troposphere 400 hPa "positive–negative" center appear alternately, the heavy rainfall region is corresponding with the strong Q-vector divergence over the negative region, rainstorm process later, the Q-vector divergence "positive–negative" alternate center weaken and disappear, the rainfall is not only related with the non-geostrophic ascending motion, the non-geostrophic sinking motion role is very important, and the non-geostrophic rising region often heavy rain area and has a good corresponding relationship; The development of the torrential rain accompanied by secondary circulation, rainfall peak periods, respectively along the longitudinal and latitudinal is a "positive–negative" center are arranged alternately, and heavy rain area is located in the south of the West and often negative area negative area value secondary circulation interchange near the ascending branch.

Keywords Rainstorm · Non-geostrophic Q-vector · Secondary circulation

---

H. Cai
Loudi Meteorological Bureau, Hunan Loudi 417000, China

J. Cai · Z. Yin (✉) · H. Liu · L. Wang
Hunan Meteorological Observatory, Changsha 410000, China
e-mail: 369543471@qq.com

H. Cai · J. Cai · Z. Yin · H. Liu · L. Wang
Hunan Key Laboratory of Meteorological Disaster Prevention and Reduction, Changsha 410007, China

C. Yuan et al. (eds.), *Proceedings of 4th International Conference on Resources and Environmental Research—ICRER 2022*, Environmental Science and Engineering,
https://doi.org/10.1007/978-3-031-31808-5_7

7.1 Introduction

Hunan Province is located on the south bank of the middle reaches of the Yangtze River, north of the Nanling Mountains. It is located between 24°38′~30°08′ N and 108°47′~114°15′ E. It belongs to the subtropical humid monsoon climate zone. Rainstorms are frequent, and persistent rainstorms often lead to floods, causing great harm to people's lives and property. Since 1978, Hoskins et al. (1978) proposed the concept of Q vector and derived the quasi-geostrophic ω equation with its divergence as the only forcing term, which intuitively shows the strength of the forcing mechanism of ω. Since the late 1980s, the quasi-geostrophic Q vector has been gradually applied to China's meteorological operations (Yang and Cao 1995; Yi et al. 1999; Shi et al. 2008; Song et al. 2009; Gao et al. 2018, 2019; Li and Li 2017; Ran et al. 2019). Subsequent studies have found that (Zhang 1999; Liu 2006; Li et al. 2002), the quasi-geostrophic Q vector is the result of assuming that the atmosphere is derived under adiabatic and quasi-geostrophic conditions, but it is not adiabatic for the actual atmosphere, especially in the rainstorm process. There is a large amount of condensation latent heat release, and the assumption of adiabatic conditions is very impractical. At the same time, due to the limitation of quasi-geostrophic approximation, it is not suitable for application in low latitudes, especially for the study of sub-synoptic scale motion. In 1998, Zhang (Zhang 1998) considered the effect of atmospheric condensation latent heat, put forward the concept of wet Q vector, and derived the expression of non-geostrophic wet Q vector and the non-geostrophic ω equation with the divergence of wet Q vector as the only forcing term from the non-adiabatic original equations, which can more truly reflect the situation of the actual atmosphere corresponding to the generation and development of rainstorm in different latitudes. In recent years, more and more meteorological scholars have begun to study the persistent heavy rain, and have achieved certain results, but the study of persistent heavy rain in Hunan using Q vector theory is still rare. In this paper, using NCEP 1° × 1° reanalysis data and conventional meteorological observation data, the Q-vector diagnostic analysis was carried out on a persistent rainstorm weather process in Hunan from June 25 to 28, 2012. The variation characteristics of the corresponding non-geostrophic wet Q-vector during the rainstorm process and the manifestation of the sub-synoptic scale system were discussed, which provided some useful reference for the future persistent rainstorm forecast.

7.2 Non-geostrophic Wet Q Vector Expression

According to Reference (Chen et al. 2009), the non-geostrophic wet Q vector is defined as Q* = (Qx*, Qy*), and the expression is:

$$Q_x* = \frac{1}{2}\left[f\left(\frac{\partial v}{\partial p}\frac{\partial u}{\partial x} - \frac{\partial u}{\partial p}\frac{\partial v}{\partial x}\right) - h\frac{\partial V}{\partial x}\nabla\theta - \frac{\partial}{\partial x}\left(\frac{LR\omega}{C_p p}\cdot\frac{\partial q_s}{\partial p}\right)\right] \tag{7.1}$$

$$Q_y* = \frac{1}{2}\left[f\left(\frac{\partial v}{\partial p}\frac{\partial u}{\partial y} - \frac{\partial u}{\partial p}\frac{\partial v}{\partial y}\right) - h\frac{\partial V}{\partial y}\nabla\theta - \frac{\partial}{\partial y}\left(\frac{LR\omega}{C_p p}\cdot\frac{\partial q_s}{\partial p}\right)\right] \tag{7.2}$$

Q_x* and Q_y* are the wet Q vector components in the x and y directions, respectively, where,

$$h = \frac{R}{p}\left(\frac{p}{1000}\right)^{R/C_p} \tag{7.3}$$

Other symbols are commonly used physical symbols. The formula of non-geostrophic wet Q vector shows that Q* depends on the difference effect of horizontal shear and vertical shear of wind, the product of horizontal gradient of wind and horizontal gradient of temperature and non-adiabatic effect. The non-geostrophic ω equation with the non-geostrophic wet Q vector divergence as the forcing term is:

$$\nabla^2(\sigma\omega) + f^2\frac{\partial^2\omega}{\partial p^2} = -2\nabla\cdot Q^* \tag{7.4}$$

$\nabla\cdot Q^*$ is the divergence of the non-geostrophic wet Q vector, assuming that the vertical motion of the atmosphere is a form of fluctuation, $\omega \propto \nabla\cdot Q^*$, when $\nabla\cdot Q^* < 0$, $\omega < 0$, ascending motion; when $\nabla\cdot Q^* > 0$, $\omega > 0$, sinking motion. In fact, it is the forcing mechanism of ascending motion that must excite the secondary circulation. The relationship between the non-geostrophic Q vector and the secondary circulation is:

$$Q_x* = \frac{1}{2}\left(f^2\frac{\partial u_a}{\partial p} - \sigma\frac{\partial\omega}{\partial x}\right) \tag{7.5}$$

$$Q_y* = \frac{1}{2}\left(f^2\frac{\partial u_a}{\partial p} - \sigma\frac{\partial\omega}{\partial y}\right) \tag{7.6}$$

It can be seen that the zonal and meridional vertical circulation is determined by the zonal and meridional components of the non-geostrophic wet Q vector, respectively. Q_x* and Q_y* always point to the updraft area and back to the downdraft area.

7.3 Process Overview and Large Scale Circulation Analysis

From 08:00 on June 25 to 08:00 on June 28, 2012 (Beijing time, the same below), a continuous rainstorm weather process occurred in the north of central Hunan. Among them, the heavy precipitation on June 26 had the widest range and the strongest intensity. The comprehensive analysis found that the heavy precipitation appeared in the northwest of Hunan, Dongting Lake area and central and western Hunan, causing serious flood disasters. Figure 7.1 shows the process precipitation distribution of 97

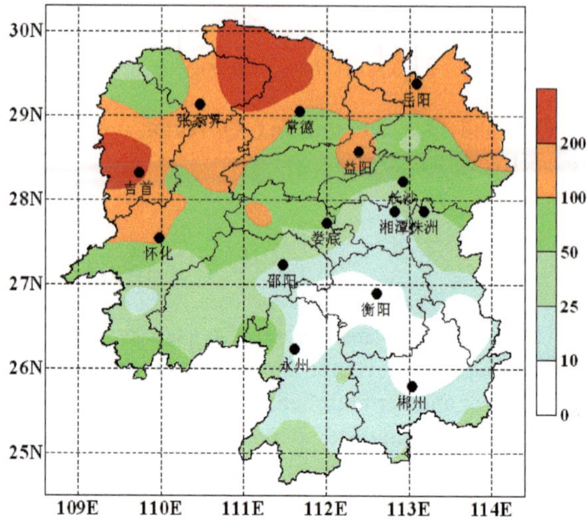

Fig. 7.1 Precipitation distribution in Hunan Province from 08:00 June 25 to 08:00 June 28, 2012 (unit: mm)

conventional meteorological observation stations in Hunan Province. As shown in the figure, the cumulative rainfall of 19 conventional meteorological observation stations in the province is 50–100 mm, 18 stations are 100–200 mm, and 6 stations are more than 200 mm. The cumulative rainfall of Cili in Zhangjiajie and Shimen in Changde is more than 300 mm, which is 329.6 mm and 318.4 mm respectively.

Figure 7.2 shows the average circulation field from 08:00 on June 25 to 08:00 on June 28, 2012. As shown in the figure, there are two ridges and one trough inverted Ω-shaped in the middle and high latitudes of Asia at 500 hPa, a high-pressure ridge from West Siberia to Northwest China, a wide inverted Ω-shaped low trough near Lake Baikal, and a high-pressure ridge from East Siberia to the Sea of Japan. The subtropical high in the low latitudes of the Western Pacific is distributed in a belt, and the position is easterly. The ridge point is located near (130° E, 22° N). The southwest monsoon is continuously strengthened during the rainstorm process. At the lower 850 hPa, the Somali cross-equatorial southwesterly airflow forms two jet centers in the range of (40° E–70° E, 5° N–15° N) and near the Bay of Bengal, respectively. The southwesterly airflow extends to the northeast to the north of South China, which transports abundant water vapor and unstable energy for the rainstorm area. The southwest vortex in Guizhou and the stable maintenance of warm shear in the middle and lower reaches of the Yangtze River are the main causes of persistent rainstorm in Hunan.

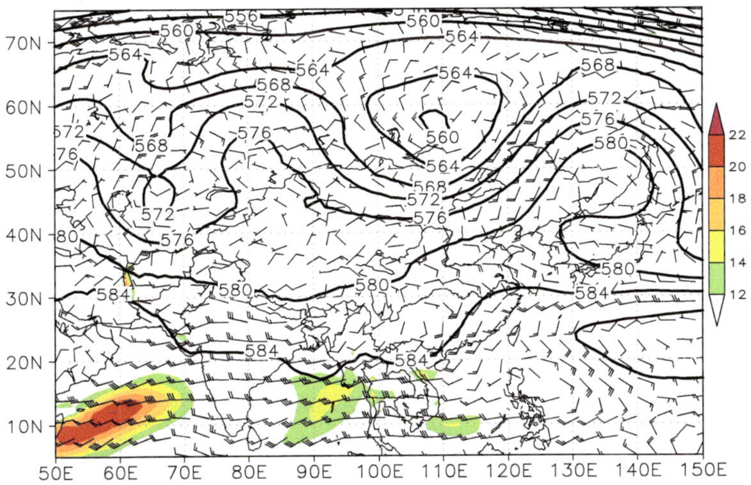

Fig. 7.2 Average 500 hPa geopotential height field from 08:00 June 25 to 08:00 June 28, 2012 (solid line, unit: dagpm), 850 hPa wind field (wind scale, unit: $m{\cdot}s^{-1}$) and torrent zone (shaded zone, unit: $m{\cdot}s^{-1}$)

7.4 Diagnostic Analysis

7.4.1 Non-geostrophic Wet Q Vector Horizontal Flow Field and Its Divergence Field Distribution

The 700 hPa non-geostrophic wet Q vector flow field and distribution during the peak period of this continuous rainstorm process were compared with the actual precipitation. As shown in Fig. 7.3, the non-geostrophic wet Q vector flow field is mainly characterized by divergence and convergence centers (convergence lines). The horizontal scale of these divergence and convergence centers (convergence lines) is about 200–300 km, with obvious mesoscale characteristics. At 20:00 on June 25 (figure omitted), there are two weak Q-vector convergence centers in Hunan. One is located at the junction of northeast Guizhou and northwest Hunan, and the other is located near Changsha-Zhuzhou-Xiangtan. The intensity of center $\nabla \cdot Q^*$ is -12×10^{-15} $hPa^{-1}{\cdot}s^{-3}$, corresponding to the next 6 h respectively at the junction of northeast Guizhou and northwest Hunan, the northern part of northern Hunan and the west of central Hunan. The location of the heavy precipitation center at the junction of northeast Guizhou and northwest Hunan is basically consistent with the position of the Q vector convergence center, while the heavy precipitation center in the northern part of northern Hunan in the next 6 h is concentrated after 21 O'clock, which cannot be predicted in the Q vector divergence field at 20 O'clock, so that the phenomenon of missing forecast occurs. In addition, the strong precipitation center in Xinhua in the west of central Hunan is not consistent with the weak Q vector convergence center near Changsha-Zhuzhou-Xiangtan, and the weak Q vector convergence center of the strong precipitation center is slightly westward. Some scholars have pointed out that (Zhang et al. 2008), the relationship between ω and proportional is obtained when ω is assumed to be a function with wavy characteristics, while ω in the actual atmosphere is not a strict wavy function. Therefore, the strongest vertical motion does not necessarily correspond to the strongest Q vector convergence zone. At 02:00 on June 26 (Fig. 3a), the weak Q-vector convergence center at the junction of northeastern Guizhou and northwestern Hunan moved toward northwestern Hunan. The convergence zone covered the entire northwestern Hunan and northern northern Hunan. The center intensity was significantly enhanced, reaching -42×10^{-15} hPa-1·s^{-3}. The Q-vector convergence center originally located near Changsha-Zhuzhou-Xiangtan weakened and tended to die out. In the next 6 h, 68 stations with precipitation greater than 50 mm and 2 stations with precipitation greater than 100 mm appeared in western Hunan, northern Huaihua, eastern Zhangjiajie and northern Changde. The heavy precipitation area was in good agreement with the strong Q-vector convergence area. At the same time, the Q vector convergence zone in northern Hunan is consistent with the falling area of heavy rainfall in the first 5 h, which makes up for the missing forecast phenomenon of heavy rainfall in northern Hunan at 20:00. At 08:00 on June 26 (Fig. 7.3b), the strong Q vector convergence zone in northwestern Hunan was further expanded, extending from southwestern Hunan to northern Hunan in a band. The original strong convergence center in northwestern Hunan was significantly weakened, and the position of the convergence center moved slightly eastward. At the same time, a weak Q vector convergence center appeared in southwestern Hunan, both centers were -12×10^{-15} hPa^{-1}·s^{-3}. In addition, at the junction of the two lakes in northern Hunan, the convergence center was significantly enhanced, reaching -30×10^{-15} hPa^{-1}·s^{-3}.

At 14:00 on 26 June (Fig. 3c), the Q vector convergence center in the northwest and north of Hunan merged into a convergence center, which was located at the junction of Zhangjiajie, Changde and Huaihua. The center value was -18×10^{-15} hPa^{-1}·s^{-3}. The evolution of this convergence center in the first 6 h was consistent with the location of heavy precipitation in the northwest of Changde, the east of Zhangjiajie and the north of Huaihua in the first 6 h of 14:00. The weak Q vector convergence center in the southwest of Hunan disappeared, and there was no heavy precipitation weather in this area within 6 h, indicating that there was a false prediction phenomenon in the weak Q vector convergence area at 08:00. In the following 6 h, the Q vector convergence center of northwestern Hunan further moved northward, and the convergence center moved to the middle of Hubei at 20:00 (figure omitted) on 26 June, with a center value of -24×10^{-15} hPa^{-1}·s^{-3}. Except for the northwest of Hunan and some areas in the north of Hunan, the other areas were divergent areas, corresponding to the weakening period of heavy precipitation in Hunan, only in the northwest of Changde and the east of Zhangjiajie. There was a small range of heavy precipitation weather, and the change trend of heavy precipitation was basically consistent with the change of Q vector convergence area in northwest Hunan. At 02:00 on 27 June (Fig. 7.3d), the position of the Q vector convergence center in the northwest of Changde did not change much, and the intensity weakened to -18×10^{-15} hPa^{-1}·s^{-3}, which was consistent with the location of the continuous heavy rainfall area in the northwest of

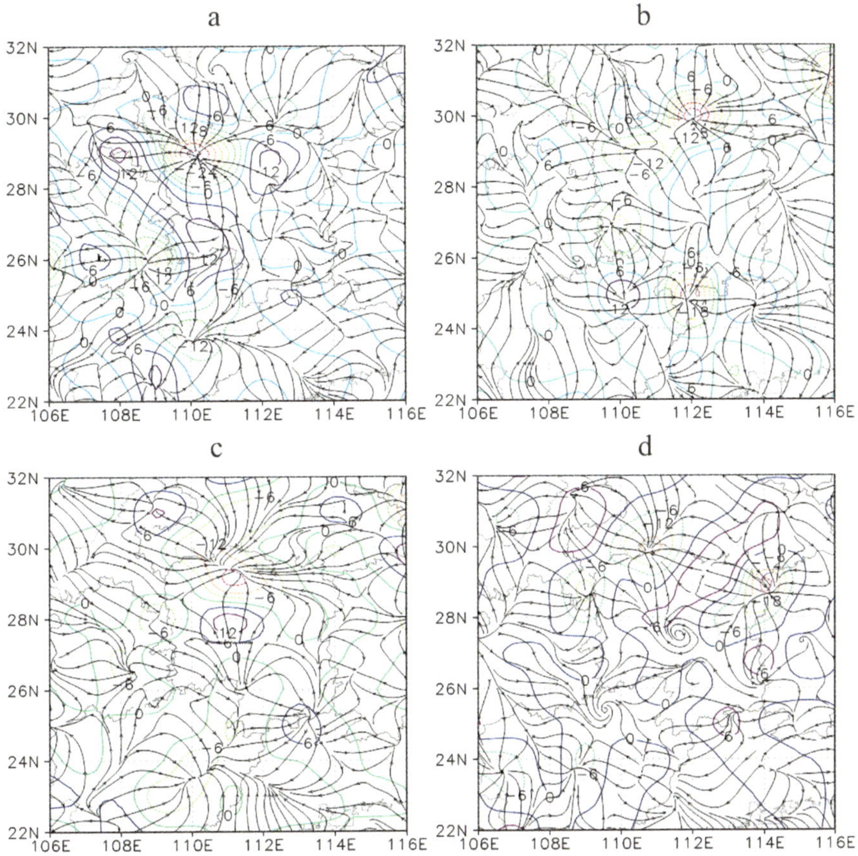

Fig. 7.3 700 hPa non-geostrophic wet Q vector flow field and distribution map (unit:10^{-15} $hPa^{-1}{\cdot}s^{-3}$) (**a** 02:00 on 26 June, **b** 08:00 on 26 June, **c** 14:00 on 26 June, **d** 02:00 on 27 June)

Changde and the east of Zhangjiajie. At the same time, a convergence center with a size of -24×10^{-15} $hPa^{-1}{\cdot}s^{-3}$ appeared in the east of Yueyang, corresponding to the short-term heavy rainfall weather in this area in the next 6 h. The above analysis shows that the change of non-geostrophic wet Q vector convergence zone has a good indication to the falling area of persistent heavy rainfall, and there is a good correspondence between the strong Q vector convergence zone and the heavy rainfall area, while the wet Q vector convergence center is stably maintained in a certain area, which is prone to persistent heavy rainfall.

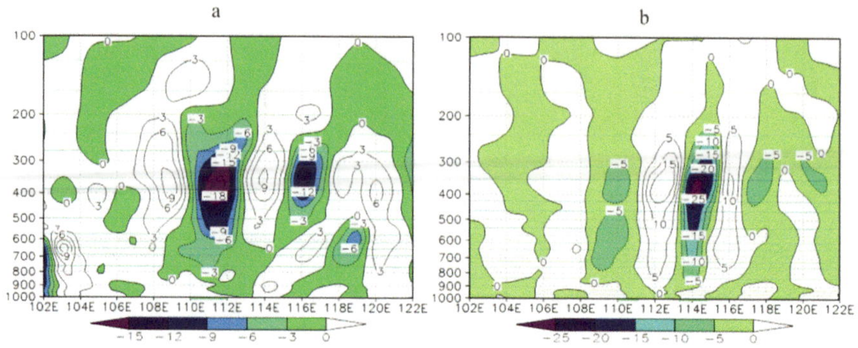

Fig. 7.4 The meridional average (28° N–30° N) zonal vertical profile (unit: 10^{-15} hPa^{-1}·s^{-3}) (**a** 14:00 on 26 June, **b** 08:00 on 27 June)

7.4.2 Vertical Distribution of Non-geostrophic Wet Q Vector Divergence

In order to reveal the relationship between $\nabla \cdot Q^*$ and heavy rainfall, the heavy rainfall area (28° N–30° N) is selected as the zonal vertical profile of the meridional average $\nabla \cdot Q^*$. Figure 7.4 shows the vertical distribution of $\nabla \cdot Q^*$ in the zonal direction during the peak period of heavy rainfall. As shown in Fig. 7.4, in the development stage of torrential rain, there are several "positive–negative" centers of $\nabla \cdot Q^*$ appearing alternately near 400 hPa, and the torrential rain area corresponds to the strong negative value area of $\nabla \cdot Q^*$. The strongest convergence area is located in the middle troposphere between 500 and 300 hPa, and the central intensity is -25×10^{-15} hPa^{-1}·s^{-3}. The negative value area corresponds to the $\nabla \cdot Q^*$ convergence area, which will force the non-geostrophic ascending motion, namely the ascending branch of the secondary circulation. At the same time, there is a positive $\nabla \cdot Q^*$ value area on the east and west sides of the negative value area, which corresponds to the non-geostrophic descending motion. After 08:00, 28 June (figure omitted), the middle troposphere along the east–west "positive–negative" alternating region disappeared, corresponding to the strong precipitation in Hunan weather tends to end. This indicates that the occurrence of torrential rain is not only related to the non-geostrophic ascending motion, but also related to the non-geostrophic descending motion, which is conducive to the formation of secondary circulation, and the negative value area often has a good correspondence with torrential rain.

7.4.3 Non-geostrophic Wet Q Vector Secondary Circulation

Studies have shown that (Qiao and He 2010) there is a close relationship between the strong convective weather and the secondary circulation. The non-geostrophic ascending and descending motions over the heavy rainfall area in Hunan are the manifestations of the secondary circulation. In order to further reveal this relationship, this paper analyzes the characteristics of the Q^* vector component and the vertical circulation in the development stage of the heavy rainfall. Figure 5a along the rainstorm area Q_x^* zonal vertical profile distribution, between 110° E and 116° E Q_x^* is "negative–positive-negative" arrangement, the rainstorm area is located in the Q_x^* negative value area on the west side of the positive and negative value of the intersection, that is, the east side of the rainstorm area Q_x^* is negative to the west, the west side of the Q_x^* is positive to the east, so there is an upward branch of the secondary circulation at the intersection of the positive and negative values on the west side of the Q_x^* negative value area, while there is a sinking branch of the secondary circulation at the intersection of the positive and negative values on the west side of the Q_x^* positive value area. In addition, the analysis of the longitudinal vertical cross section of Q_y^* along the torrential rain area (Fig. 7.5b) shows that there is also a meridional secondary circulation circle over the torrential rain area. The torrential rain area is located at the intersection of the positive and negative values on the south side of the negative Q_y^* value area, corresponding to the ascending branch of the secondary circulation, while the intersection of the positive and negative values on the south side of the positive Q_y^* value area is the descending branch of the secondary circulation. The above conclusions can be confirmed by the vertical circulation field. The 110° E–111° E, 114° E and 28° N–29° N near the intersection of the positive and negative values on the west side of the Q_x^*-negative area and the south side of the Q_y^*-negative area are updrafts, and the 113° E and 28° N near the intersection of the positive and negative values on the west side of the Q_x^*-positive area and the south side of the Q_y^*-positive area are downdrafts. Therefore, the secondary circulation has obvious indicative function in forecasting rainstorm and its falling area, and the analysis of non-geostrophic wet Q vector can provide a very important reference for rainstorm forecasting.

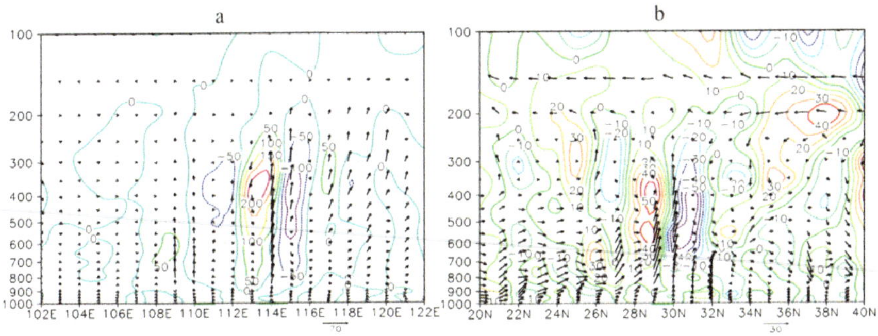

Fig. 7.5 Meridional mean (28° N–30° N) and latitudinal vertical profile (u, −50 ω) at 08:00 on 27 June (**a**), zonal mean (109° E–114° E) and meridional vertical profile (v, −50 ω) at 08:00 on 27 June (**b**) (Contour, unit: 10^{-11} m·hPa^{-1}·s^{-3})

7.5 Conclusions

In this paper, a continuous rainstorm weather process occurred in Hunan on June 25–28, 2012 is analyzed. The continuous precipitation is mainly caused by the continuous upward motion, and the strong diabatic heating is mainly caused by the strong water vapor convergence in the middle and lower layers. The relationship between non-geostrophic wet Q vector and secondary circulation and precipitation is analyzed, and the following conclusions are obtained:

(1) The change of the non-geostrophic wet Q vector convergence zone in the middle and lower levels has a good indication to the heavy rainfall area. The strong Q vector convergence center is conducive to the occurrence of heavy rainfall, and the strong Q vector convergence center is stably maintained in a certain area, which is prone to persistent heavy rainfall.

(2) There are multiple $\nabla \cdot Q^*$ "positive–negative" centers appearing alternately near 400 hPa in the middle troposphere. The rainstorm area corresponds to the strong $\nabla \cdot Q^*$ negative area, and the non-geostrophic ascending motion is forced in the Q^* convergence area. The $\nabla \cdot Q^*$ positive area on the east and west sides of the negative area corresponds to the non-geostrophic sinking motion. The rainstorm is not only related to the non-geostrophic ascending motion, but also the non-geostrophic sinking motion.

(3) The W vector component can well reveal the spatial distribution characteristics of the secondary circulation. During the peak period of the torrential rain, X and Y are alternately arranged in multiple "positive–negative" centers along the latitudinal and longitudinal directions, respectively. The torrential rain area is often located near the ascending branch of the secondary circulation at the intersection of the positive and negative values on the west side of the X negative area and the south side of the Y negative area.

Acknowledgements The paper is funded by the following project: China Meteorological Administration Forecaster Special Project (CMAYBY2020-085); Key Projects of Hunan Meteorological Bureau (XQKJ22A004); Key Fields Research and Development Project of Hunan Provincial Department of Science and Technology (2019SK2161); Hunan Business Capacity Building (NLJS14); Key Projects of Hunan Meteorological Bureau (XQKJ19B016).

References

Chen HZ, Zhang DS, Luo ZR (2009) Diagnostic analysis of heavy rainfall in north Hunan Province. Torrential Rain Disasters 28(4):373–376

Gao LX, Liu C, Yang XX et al (2018) Ageostrophic wet Q-vector analysis of a persistent torrential rainfall event in the northwest of Shandong Province. J Mar Meteorol 38(3):75–82

Gao S, Xiao TG, Wu LJ (2019) Diagnostic analysis of wet Q-vector for a rainstorm process in the middle and lower reaches of Yangtze River. Meteorol Environ Sci 42(3):58–67

Hoskins BJ, Draghici I, Davies HC (1978) A new look at the ω-equation. Quart J Roy Meter Soc 104(11):31–38

Li SS, Li GP (2017) Diagnostic analysis based on wet Q-vector of a shear line with rain on the east side of Qinghai-Tibetan Plateau under the saddle pattern circulation background field. Plateau Meteorol 36(2):317–329

Li Y, Zhang TF, Guo FR (2002) Ageostrophic wet Q-vector analysis of a persistent rainstorm process in Yunnan. J Nanjing Inst Meteorol 25(2):259–264

Liu YC (2006) Wet Q-vector analysis of a continued heavy rain process over the eastern Qinghai-Xizang plateau. Meteor Mon 32(12):43–49

Qiao CG, He Z (2010) Ageostrophic wet Q-vector analysis of a rainstorm process in Henan. Meteorol Environ Sci 33(2):48–52

Ran LK, Li Z, Zhang YB et al (2019) The diagnostic analysis of Q vector during a heavy rain event in north China. Torrential Rain Disasters 38(1):17–30

Shi WZ, Zhang PP, Wu T et al (2008) Contrastive analysis of heavy rain cases in Hubei province. Torrential Rain Disasters 27(3)219–224

Song QZ, Qiao CG, Gao YY (2009) Synoptic analysis of the super rainstorm process in Henan. Meteorol Environ Sci 32(3):6–9

Yang XY, Cao XX (1995) An analysis of the relationship between quasi-geostrophic Q-vector and heavy rain. Meteor Mon 21(11):17–21

Yi Y, Duan X, Pan L (1999) A quasi-geostraphic Q-vector analysis of Kunming quasi-stationary front. Meteor Mon 25(8):6–10

Zhang XW (1998) An expression of the wet Q-vector and application. Meteor Mon 24(8):3–7

Zhang X (1999) The expression of the modified Q-vector and its application. J Trop Meteorol 15(2):162–167

Zhang GZ, Li G, Bai JH et al (2008) Role of coupled jets at different levels in Henan regional heavy rain during 12–14 July 2007. Meteorol Environ Sci 31(2):7–12

Chapter 8
Analysis on Characteristics of Lightning Activities in Changsha from 2011 to 2020

Shaojuan Wang and Daoping Wang

Abstract Based on the lightning location data and 30 m-precision elevation data from 2011 to 2020 in Changsha, the characteristics of cloud-to-ground (CG) lightning activities were analyzed in different terrain conditions in Changsha by mathematical statistics. The results show that CG lightning mainly occurs in the plains and hills, and the current intensity is mainly concentrated in 10–100 kA. With the increase of altitude, the frequency of CG lightning and the proportion of small amplitude CG lightning current decrease, while the proportion of large amplitude CG lightning current increases. Spring and summer are the high incidence seasons of lightning activities in each terrain, while winter is the active period of positive CG lightning. There are many CG lightning activities from afternoon to evening in the day. Among them, the CG lightning activities are the strongest at 16:00 in the plains and hills, and the CG lightning activities are the strongest at 15:00 in the mountain. From the seasonal distribution of lightning current steepness, the variation range of positive and negative CG lightning steepness in the plains and hills is smaller than that in the mountain. During the day, with the increase of altitude, the minimum value of lightning current steepness decreases and the maximum value of lightning current steepness increases.

Keywords Cloud-to-ground lightning frequency · Lightning current intensity · Elevation data · Terrain

---

S. Wang (✉) · D. Wang
Hunan Meteorological Disaster Prevention Technology Center, Changsha 410007, China
e-mail: 1067657725@qq.com

Key Laboratory of Meteorologicol Disaster Prevention and Mitigation in Hunan Province, Changsha 410118, China

C. Yuan et al. (eds.), *Proceedings of 4th International Conference on Resources and Environmental Research—ICRER 2022*, Environmental Science and Engineering,
https://doi.org/10.1007/978-3-031-31808-5_8

8.1 Introduction

Lightning is a discharge phenomenon that occurs in the atmosphere. It has the characteristics of long distance, short time and large energy. With the rapid development of economy and the popularization of electronic and electrical equipment, the threat of lightning to human beings is becoming more and more serious, which has been identified as one of the top ten natural disasters by the United Nations (Ming et al. 2008a). The CG lightning activities is greatly affected by the underlying surface factors such as region and terrain, and has strong local and regional characteristics (Xiushu et al. 2003; Xiushu and Toumi 2003; Xueliang et al. 2019; Shenghao and Lei 2016). Therefore, studying the characteristics of CG lightning activities in different regions is of great significance for further studying the physical characteristics of lightning and improving the defense ability of lightning disasters in different regions.

Since the beginning of the twenty-first century, scholars have carried out a series of studies on the law of lightning activities. Xueliang Wang, Jiaqi Li et al. analyzed the characteristics of CG lightning activities in Hubei Province from 2006 to 2009 in terms of spatial distribution and temporal distribution by using Advanced TOA and Direction system (ADTD) data (Xueliang et al. 2010; Jiaqi et al. 2011). Zhongzhi Fan et al. analyzed the topographic factors affecting CG lightning activities by using lightning location data (Zhongzhi and Yating 2017). Xuedong Cui et al. analyzed the relationship between terrain and thunderstorm path in Zhejiang Province and found that the main path of thunderstorm was hilly terrain, with fewer plains and mountains, and the higher the terrain, the fewer paths (Xuedong and Weibin 2021); Ming Ma, Yi Zhang, Qingfeng Zeng et al. studied the distribution characteristics and causes of lightning disasters by using the statistical data of lightning disasters for many years, and provided reference for the development of disaster prevention and mitigation (Ming et al. 2008b; Yi et al. 2018; Qingfeng et al. 2017). Qiuyan He, Yanting Tian, Ankun Wu et al. from the perspective of vulnerability, selected CG lightning strike density, frequency of lightning disaster, economic and life vulnerability modulus and other evaluation indicators to carry out lightning disaster risk zoning analysis (Qiuyan et al. 2018; Yanting et al. 2012; Ankui 2018).

China has a vast territory. Studying the characteristics of CG lightning activities in different regions can not only provide a basis for the development of lightning protection, but also improve the defense ability of lightning disasters and prevent or reduce the occurrence of lightning disasters. Therefore, this paper uses CG lightning location data and Digital Elevation Model (DEM) data from 2011 to 2020 in Changsha to analyze the characteristics of CG lightning activities, in order to provide reference for lightning protection design and disaster prevention and mitigation in Changsha.

8.2 Data and Methods

The CG lightning location data comes from the lightning location system of Hunan Province, which is an ADTD developed, produced and distributed by the Scientific Research Center of Chinese Academy of Sciences and Beijing Huayun Oriental Detection Technology Co., Ltd., and completed in the whole province in early 2008. The system consists of 10 detection sub-stations located in Changsha, Yiyang, Yueyang, Changde, Zhangjiajie, Huaihua, Shaoyang, Yongzhou, Chenzhou and Hengyang respectively. It is mainly used to detect CG lightning. The detection efficiency is ±80%, and the parameters such as the occurrence time of CG lightning, latitude and longitude position, lightning current polarity, peak intensity, and lightning current rising steepness can be obtained. The time accuracy can reach 10–7 s, and the positioning accuracy can reach 300 m.

The DEM elevation data comes from the geographic data cloud platform, the data type is image, and the spatial resolution is 30 m. According to the terrain division standard, the terrain of Changsha is divided into plains (altitude < 200), hills ($200 \leq$ altitude < 500), low mountain ($500 \leq$ altitude < 1000) and high mountain (altitude ≥ 1000), and the contour map of altitude is drawn (Fig. 8.1).

Considering that the CG lightning current monitoring range of the ADTD is (±1 kA, ±500 kA), and there is a small amplitude of cloud lightning interference, therefore, the CG lightning current with intensity greater than 500 kA is deleted in this study, and the CG lightning current data of −5 to 5 kA is eliminated at the same. The CG lightning current wave front steepness interval is less than 500 kA/μs and greater than 0 kA/μs.

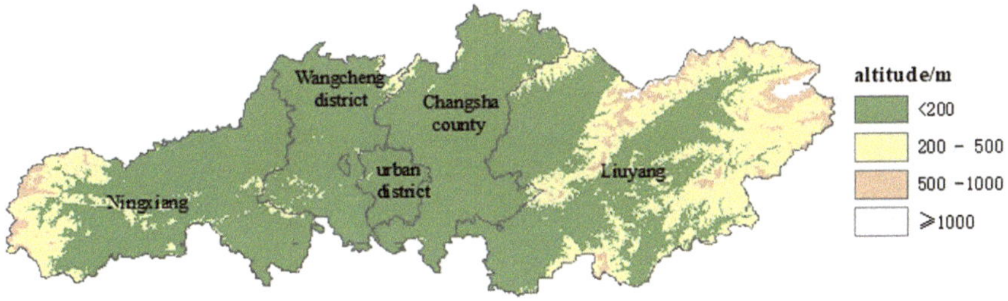

Fig. 8.1 Schematic diagram of altitude of Changsha

8.3 Results and Analysis

8.3.1 *Distribution Characteristics of CG Lightning Frequency*

The spatial distribution characteristics of CG lightning activities are analyzed with a grid size of 1 km × 1 km for Kriging interpolation from 2011 to 2020 in Changsha. It can be seen from Fig. 8.2 that the western of Ningxiang is the most active area of CG lightning activities, and the CG lightning density in the central and eastern regions mainly falls below 2.6 times/km^2 a, showing the regional distribution characteristics of more in the west and less in the east, less in the south and more in the north. According to the statistics of the proportion of total CG lightning frequency in each terrain area in Changsha, it can be seen that the plain terrain accounts for the largest proportion, accounting for 74.82%, followed by hills, accounting for 20.29%, and the proportion of CG in low mountain and high mountain is 4.50% and 0.39% respectively.

A large number of studies have shown that, the distribution of CG lightning activities is related to topography and landforms (Xueliang et al. 2020; Minfeng et al. 2000). In order to further study the spatial distribution difference of CG lightning density in Changsha, the number of CG lightning in each altitude interval was counted at an altitude of 100 m from 2011 to 2020, as shown in Fig. 8.3. It can be seen from Fig. 8.3 that the CG lightning frequency increases with fluctuation from 1.227 to 1.2447 times km^{-2} a^{-1} when the altitude from 200 to 700 m. The frequency of CG lightning decreases from 1.2447 to 0.896 times km^{-2} a^{-1} with altitude from 700 to 1550 m. Statistical analysis shows that the frequency of CG lightning decreases with the increase of altitude, which is basically consistent with the conclusion of Reference (Xueliang et al. 2020). According to the research results obtained by relevant researchers, the concentration of aerosol ions in the atmosphere affects the occurrence of lightning activities. According to the research results obtained by meteorological observations, aerosols in the atmosphere are mostly distributed in the troposphere below 5 km altitude, and the solubility decreases exponentially with

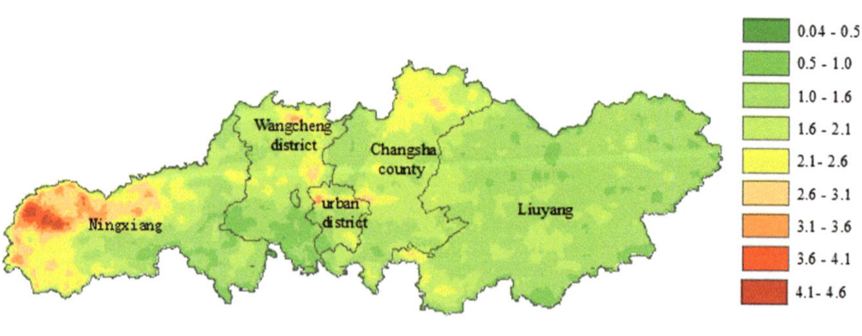

Fig. 8.2 Average annual density of CG lightning from 2011 to 2020 in Changsha

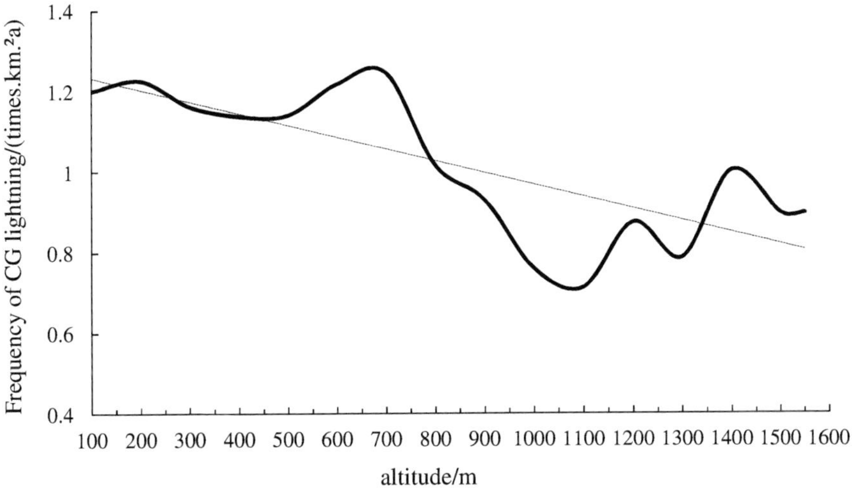

Fig. 8.3 Frequency of CG lightning varies with altitude from 2011 to 2020 in Changsha

the increase of altitude, so that the frequency of CG lightning decreases with the increase of altitude (Weimin 2006).

8.3.2 Distribution Characteristics of CG Current Intensity

The absolute value of CG lightning intensity on different terrains is analyzed from 2011 to 2020 in Changsha (Table 8.1). It can be seen that the absolute value of CG lightning intensity is mainly concentrated in 10–100 kA, accounting for 95.8% (whole area), 96.17% (plains), 95.07% (hills), 94.05% (low mountain), 89.82% (high mountain), and the average value of CG lightning intensity above 200 kA is less than 1%.

Table 8.1 Statistics on the division of absolute value of CG lightning intensity under different terrain conditions from 2011 to 2020 in Changsha

Terrain	CG lightning intensity/kA						
	(0, 10]	(10, 20]	(20, 40]	(40, 60]	(60, 100]	(100, 200]	(200, 500]
Whole area (%)	0.24	7.81	48.41	27.40	12.18	3.57	0.38
Plains (%)	0.33	8.39	50.25	26.17	11.36	3.17	0.33
Hills (%)	0.06	6.96	44.89	29.87	13.35	4.41	0.45
Low mountain (%)	0.02	7.16	41.32	30.01	15.56	5.24	0.69
High mountain (%)	0.00	1.53	25.64	37.69	24.96	8.15	2.04

According to the maximum CG lightning current of 100 kA and the minimum CG lightning current of 15.8 kA given in the code for Design Protection of structures against lightning (GB50057-2010), we define the CG lightning current ≤20 kA as small amplitude CG lightning current, and the CG lightning current >100 kA as large amplitude CG lightning current. Comparing the proportion of small amplitude CG lightning current on the four terrains, it can be seen that the proportion of small amplitude CG lightning current in the plain area is the highest, 8.72%, and the high mountain is the least, 1.53%, that is, the proportion of small amplitude CG lightning current in the plain area is 5.7 times that of the high mountain; The proportion of large-amplitude lightning currents increases with altitude, and the proportion of high mountain area is 6.2 times that of plain. It can be seen that the probability of small amplitude lightning current shielding failure in low-altitude plain is higher than that in high-altitude mountain area, and high-altitude mountainous area is more likely to have the risk of large-amplitude lightning current counter attack than plains and hilly areas. The main reason is that the smoke and dust emitted by human activities such as urban pollution and automobile exhaust decreases with the increase of altitude, and then the charged ions in the air also decrease relatively, which adds the amplitude of lightning current of thunderstorm cloud to ground discharge. Therefore, the proportion of large amplitude CG lightning current is related to the concentration of charged ions in the air.

8.3.3 Time Variation Law of CG Lightning Activities

Seasonal Variation. Studies have shown that the occurrence of CG lightning varies greatly depending on different climatic backgrounds, especially the typical climatic events with periodic and regional characteristics have a particularly obvious impact on regional lightning activities (Wei et al. 2013; Cuimei et al. 2014). According to the division method of four seasons, the CG lightning data of different terrains are analyzed by seasons in Changsha from 2011 to 2020, with March to May as spring, June to August as summer, September to November as autumn, and December to February as winter, the results are shown in Table 8.2. It can be seen from the table that the frequency of total CG lightning in the plains, hills, low mountain and high mountain has significant variation characteristics with seasons, the proportion of total CG lightning is the largest in summer, which is 70.4%, 72.57%, 72.22% and 64.57% respectively, and more in spring than autumn, at least in winter, at least in winter, which is consistent with the statistical results of literature (Qiuyan et al. 2019). The total CG lightning in spring and summer accounts for about 90% in each terrain, which shows that spring and summer are the high incidence seasons of lightning activities, and the prevention of lightning disasters should be strengthened.

From the data in Table 8.2, it can be found that the proportion of positive CG lightning is the lowest in summer on plains, hills, low mountain and high mountain, which is 2.58% (plains), 2.92% (hills), 3.13% (low mountain) and 2.11% (high mountain) respectively. The highest proportion is winter, which is about 28 times

Table 8.2 Seasonal distribution of total and positive CG lightning under different terrain conditions from 2011 to 2020 in Changsha

Terrain		Season			
		Spring (%)	Summer (%)	Autumn (%)	Winter (%)
Plains	Total CG lightning proportion of year	20.85	70.4	8.25	0.5
	Proportion of positive CG lightning to total CG lightning	9.28	2.58	2.59	29.98
Hills	Total CG lightning proportion of year	19.32	72.57	7.58	0.54
	Proportion of positive CG lightning to total CG lightning	9.22	2.92	4.22	31.06
Low mountain	Total CG lightning proportion of year	19.96	72.22	7.30	0.52
	Proportion of positive CG lightning to total CG lightning	10.36	3.13	6.57	20
High mountain	Total CG lightning proportion of year	25.21	64.57	9.37	0.85
	Proportion of positive CG lightning to total CG lightning	12.16	2.11	7.27	60

higher than that in summer (high mountain), followed by spring, indicating that winter is an active period of positive CG lightning. The main reason is that the warm cloud area of thunderstorm cloud is thicker in summer, and the liquid water content is less in thunderstorm cloud, which makes the concentration of large particles in the cloud smaller. In addition, the reverse temperature layer is lower, and the range of positive charge area in the lower part is smaller and the intensity is weaker, making the negative charge region is stronger than the lower positive charge region. Therefore, positive CG lightning is less in summer.

Daily Variation. The analysis of the characteristics of daily variation of the CG lightning frequency in the plains, hills, low mountain and high mountain from 2011 to 2020 (Fig. 8.4) shows that the daily variation characteristics of the proportion of CG lightning frequency in each terrain are obvious, the temperature is relatively low from morning to morning, and the proportion of CG lightning frequency is relatively low. The temperature is relatively high from afternoon to evening, and the proportion of CG lightning frequency is also relatively high. Mainly because of the influence of the afternoon solar radiation, the ground temperature rises, the thermal conditions are good, and it is easy to develop thunderstorm activities (Tie and Xiushu 2004). The statistics show that the CG lightning activities is the strongest at 4 p.m on the plains and hills, accounting for 15.22% and 11.44% respectively, and the CG lightning activities is the strongest at 3 p.m on the low mountain and high mountain, accounting for 11.58% and 17.69% respectively. The maximum value of plains and hills appears 1 h later than that of low mountain and high mountain. The proportion of CG lightning is higher in plains and hills than that in mountain at 12–21 h, and the proportion of CG lightning is higher in mountain than that in plains and hills at 0–11 h and 22–23 h.

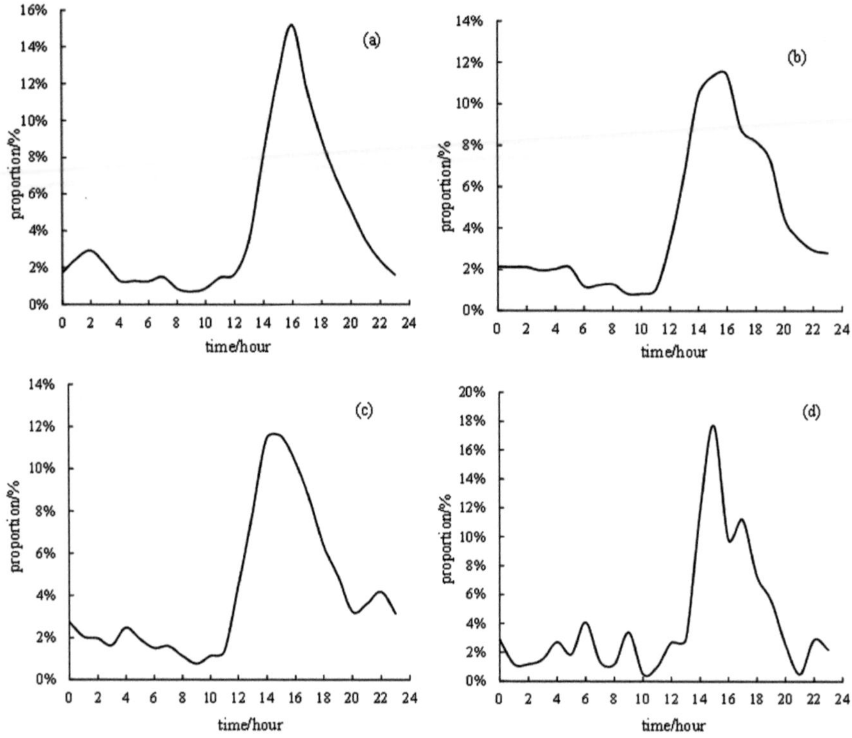

Fig. 8.4 Distribution of the next-day variation of CG lightning frequency over each terrains from 2011 to 2020 (**a** plains **b** hills **c** low mountain **d** high mountain)

8.3.4 Distribution Characteristics of Lightning Steepness

Seasonal Variation. The data of lightning current wave head steepness in Changsha from 2011 to 2020 is analyzed (Table 8.3). The positive CG lightning steepness is 12.98–15.93 kA/μs, the negative CG lightning steepness is 10.98–13.17 kA/μs, and the difference between positive and negative CG lightning steepness is 2–3.31 kA/μs. The positive CG lightning steepness is 12.08–16.87 kA/μs, the negative CG lightning steepness is 11.61–14.96 kA/μs, and the difference between the positive and negative CG lightning steepness is −2.88 to 3.16 kA/μs. The positive CG lightning steepness is 11.22–25.38 kA/μs, the negative CG lightning steepness is 11.09–15.35 kA/μs, and the difference between positive and negative CG lightning steepness is −2.48 to 10.03 kA/μs. The steepness of positive and negative CG lightning are 8.23–21.14 kA/μs and 11.43–20.6 kA/μs, and the difference between positive and negative CG lightning is −12.37 to 5.6 kA/μs. It can be seen that the variation range of positive and negative CG lightning steepness in plains and hills is smaller than that in low mountain and high mountain.

Table 8.3 Seasonal distribution of CG lightning current wave front steepness under different terrain conditions in Changsha from 2011 to 2020

		Season			
		Spring	Summer	Autumn	Winter
Plains	Positive CG lightning	12.98	15.13	15.93	15.22
	Negative CG lightning	10.98	12.50	12.62	13.17
	Differential value	2.00	2.63	3.31	2.05
Hills	Positive CG lightning	12.80	15.22	16.87	12.08
	Negative CG lightning	11.61	12.06	13.75	14.96
	Differential value	1.19	3.16	3.12	−2.88
Low mountain	Positive CG lightning	11.22	14.26	25.38	11.33
	Negative CG lightning	11.09	11.78	15.35	13.81
	Differential value	0.13	2.48	10.03	−2.48
High mountain	Positive CG lightning	17.03	11.00	21.14	8.23
	Negative CG lightning	11.43	13.87	16.84	20.60
	Differential value	5.60	−2.87	4.30	−12.37

Because the CG lightning current steepness is proportional to the induced voltage generated by the CG lightning current, that is, the greater the CG lightning current steepness is, the higher the CG lightning current induced voltage is, and the greater the lightning hazard is. Comparing the positive and negative CG lightning steepness in plains, hills, low mountain and high mountain in Table 8.3, it can be seen that the positive CG lightning in plain is more harmful than the negative CG lightning, and the positive CG lightning in autumn is the most harmful, negative CG lightning is the most harmful in winter. The lightning hazard of positive CG lightning in spring, summer and autumn is greater than that of negative CG lightning. In contrast, the lightning hazard of positive CG lightning is the greatest in autumn, and the lightning hazard of negative CG lightning is the greatest in winter. The lightning hazard of positive CG lightning is greater than that of negative CG lightning in spring, summer and autumn in low mountain, while the opposite is true in winter. The lightning hazard of positive CG lightning in spring and autumn is greater than that of negative CG lightning, while in summer and winter, the lightning hazard of positive CG lightning is the largest in autumn, and the lightning hazard of negative CG lightning is the largest in winter.

Daily Variation. It can be seen from the diurnal variation distribution of CG lightning steepness in Fig. 8.5 that the diurnal variation of CG lightning steepness increases with the increase of altitude. The minimum value of CG lightning steepness decreases with the increase of altitude, which is 10.42 kA/μs (plains), 10.33 kA/μs (hills), 10.09 kA/μs (low mountain), 9.23 kA/μs (high mountain). The maximum value of CG lightning steepness increases with the increase of altitude, which is 13.44 kA/μs (plains), 14.55 kA/μs (hills), 15.2 kA/μs (low mountain), 20.53 kA/μs (high mountain). Statistics show that the difference of CG lightning gradient

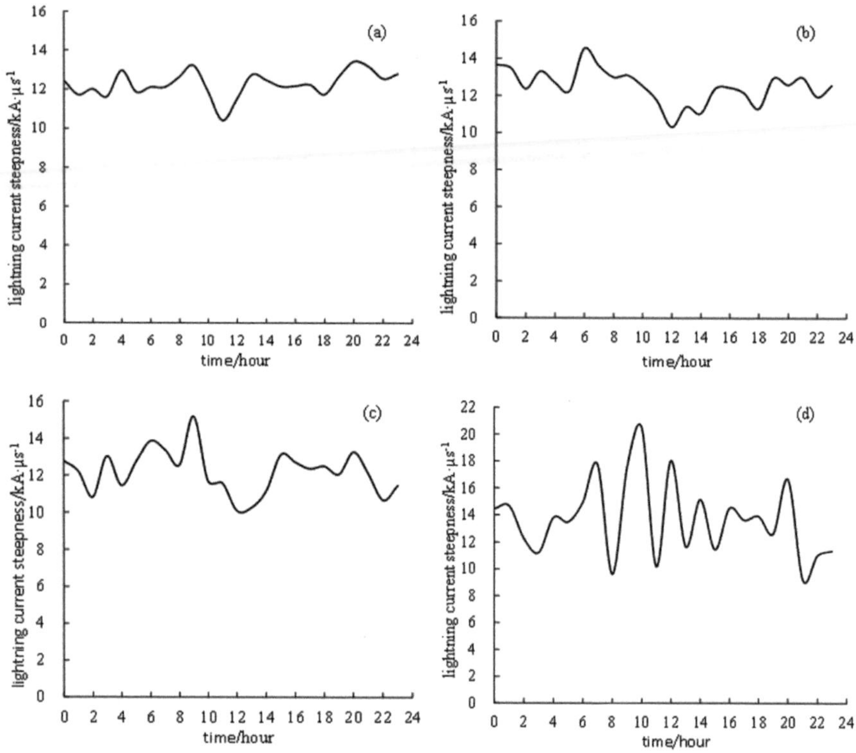

Fig. 8.5 Diurnal variation distribution of lightning current steepness over various terrains from 2011 to 2020 (**a** plains **b** hills **c** low mountain **d** high mountain)

between daytime and nighttime is small in plains, hills, low mountain and high mountain, except for high mountain (1.18 kA/μs), others are below 1 kA/μs; the CG lightning steepness at 5–10 o'clock in each terrain is greater than that in other periods, indicating that the lightning induction hazard at 5–10 o'clock in plains, hills, low mountain and high mountain is greater than that in other periods.

8.4 Conclusion

Using CG lightning location data and DEM data in Changsha from 2011 to 2020, the characteristics of CG lightning activities are analyzed in Changsha, and the following conclusions are obtained:

(1) From 2011 to 2020, CG lightning density is generally characterized by the distribution of more in the west and less in the east, less in the south and more in the north in Changsha. The total frequency of CG lightning is the largest in

the plains, followed by the hills, and the lowest in the mountain. The frequency of CG lightning decreases with the increase of altitude.

(2) The CG lightning intensity is mainly concentrated in 10–100 kA in Changsha from 2011 to 2020. The proportion of small CG lightning is the highest on the plain, which is 8.72%, and the lowest on the mountain, which is 1.53%. The proportion of large CG lightning current increases with the increase of altitude, and the proportion on the mountain is 6.2 times than that on the plain.

(3) The proportion of total CG lightning on plains, hills, low mountain and high mountain is the highest in summer, accounting for 70.4%, 72.57%, 72.22% and 64.57% respectively, spring is more than autumn, and winter is the least. The positive CG lightning is active in winter, and the positive CG lightning is about 28 times higher than that in summer (high mountain). In a day, the frequency of CG lightning is relatively high from afternoon to evening. The proportion of lightning is higher in plains and hills than that in mountain at 12–21 h, and the proportion of lightning is higher in mountain than that in plains and hills at 0–11 h and 22–23 h.

(4) From the seasonal distribution of CG lightning current steepness, the variation range of positive and negative CG lightning steepness on plains and hills is smaller than that on low mountain and high mountain. In a day, the minimum value of CG lightning steepness on plains, hills, low mountain and high mountain decreases with the increase of altitude, which is 10.42 kA/μs (plains), 10.33 kA/μs (hills), 10.09 kA/μs (low mountain), 9.23 kA/μs (high mountain). The maximum value of CG lightning gradient increases with the increase of altitude, which is 13.44 kA/μs (plains), 14.55 kA/μs (hills), 15.2 kA/μs (low mountain), 20.53 kA/μs (high mountain); the CG lightning steepness at 5–10 o'clock on plains, hills, low mountain and high mountain is greater than that at other times.

References

Ankui W (2018) Risk assessment and zoning of lightning disaster in Guizhou province. China J Agric Resour Reg Plan 39(2):88–93

Cuimei S, Xingjian S, Peng Q et al (2014) Analysis of climatic characteristics and synoptic forecasting method of thunderstorm in Zhenjiang. J Meteorol Environ 37(3):8–13

Jiaqi L, Shuanghe S, Baicheng X et al (2011) Analysis of lightning frequency distribution characteristics based on ADTD System. J Trop Meteorol 27(5):710–716

Minfeng Z, Xinsheng L, Zhengmo G (2000) The features of lightning flash activities in North of China. Plateau Meteorol 19(3):7–284

Ming M, Weitao L, Yijun Z et al (2008a) Characteristics of lightning exposure in China from 1997 to 2006. J Appl Meteorol Sci 19(4):93–400

Ming M, Weitao L, Yijun Z et al (2008b) Analysis of lightning disaster in china and their correlative factors. Adv Earth Sci 223(8):856–865

Qingfeng Z, Dongpu X, Yin J et al (2017) Analysis of lightning activities and characteristics of lightning disaster in Shenzhen from 2013 to 2016. Guangdong Meteorol 39(6):36–39

Qiuyan H, Bin G, Jiayan Y et al (2018) Characteristics of thunderstorm and disaster risk regionalization in Changsha. Hubei Agric Sci 57(23):65–70

Qiuyan H, Yueyu L, Jiayan Y et al (2019) Analysis of lightning activities and lightning disaster characteristics in Hunan province. Adv Meteorol Sci Technol 9(6):44–48

Shenghao Z, Lei Y (2016) Relationship between land cover and cloud-to-ground lightning distribution based on MODIS data. Meteorol Sci Technol 44(5):822–827

Tie Y, Xiushu Q (2004) Spatial and temporal distribution characteristics of lightning activities in China from satellite observation. Plateau Meteorol 23(4):488–494

Wei Z, Hangwei T, Jun Z et al (2013) Temporal and spatial distribution characteristics and influencing factors of lightning in Zhejiang province. Power Syst Technol 37(5):1425–1431

Weimin C (2006) Lightning principle, 2nd edn. Meteorological Press, Beijing

Xiushu Q, Junjun Z, Tie Y (2003) Global lightning activities and their regional differences observed from the satellite. Chin J Geophys 46(6):743–750

Xiushu Q, Toumi R, Junjun Z (2003) Characteristics of lightning activities over the Central Tibetan Plateau and its response to the maximum unstable energy of convection. Chin Sci Bull 48(01):89–92

Xuedong C, Weibin Z (2021) Spatiotemporal characteristics of thunderstorm tracks in Zhejiang Province from 2007 to 2017. J Meteorol Environ 37(1):59–66

Xueliang W, Xuechun L, Xiaoyan H et al (2010) Analysis of spatial and temporal distribution characteristics of cloud-to-ground lightning in Hubei area. Meteorol Mon 36(10):91–96

Xueliang W, Tianye Y, Shan H (2019) Distribution characteristics and comparative analysis of lightning parameters between lake and land. J Meteorol Environ 35(6):132–138

Xueliang W, Tianye Y, Shan H et al (2020) Influence of regional altitude on the distribution of cloud-to-ground lightning parameters. High Volt Eng 46(4):1206–1215

Yanting T, Mengheng W, Fengqi S et al (2012) Evaluation and regionalization of lightning disaster vulnerability over Hebei province. Meteorol Sci Technol 40(3):507–512

Yi Z, Hao L, Xuewen B (2018) Statistical analysis of lightning disaster characteristics in Zhejiang province from 2007 to 2013. J Meteorol Environ 41(2):139–143

Zhongzhi F, Yating F (2017) Spatiotemporal characteristics of lightning activities and its correlation with terrain in Shanxi Province. Guangdong Meteorol 39(4):69–72

Chapter 9
Research on the Relationship Between Geographical Conditions Monitoring Data and Land Surface Temperature

Li Li, Fengmin Wu, and Zhipeng Zheng

Abstract It is of great significance for analysis between land cover and land surface temperature (LST) which could improve the urban thermal environment. In this paper, we obtained the LST by the Radiative Transfer Equation method based on Landsat-8 data in 2021. The relationship between LST and land cover was analyzed by rasterizing to a grid size of 100 m × 100 m. The landscape index was obtained to analyze the characteristics of LST under different landscape patterns such as patch density, largest patch index, edge density, Shannon's diversity index and aggregation index. The result showed, the cooling effect of the water area was obvious with 9° lower than building(s) and structure which had a negative correlation with LST, however bare surface, planting and vegetation had a low correlation. The edge complexity of different land cover types seemed quite different with high plaque fragmentation. The largest aggregation index was highest in water area and higher in forest and grass areas. The edge density of forest and grass was largest. All the landscape indexes had a certain correlation with LST.

Keywords Landsat-8 data · Land surface temperature · Landscape index

9.1 Introduction

The agglomeration of population and large-scale construction of cities resulted in high land surface temperature (LST) that caused bad effects on urban ecological environment and human life (Zhao 2008; Qiu et al. 2008). Researches on the influence between land cover and LST were significant for improving the urban thermal environment and guiding urban planning and construction in the future (Cui et al.

L. Li
Chongqing Vocational Institute of Engineering, Chongqing, China

F. Wu (✉) · Z. Zheng
Chongqing Geomatics and Remote Sensing Center, Chongqing, China
e-mail: wfmrice@163.com

C. Yuan et al. (eds.), *Proceedings of 4th International Conference on Resources and Environmental Research—ICRER 2022*, Environmental Science and Engineering,
https://doi.org/10.1007/978-3-031-31808-5_9

2018). The methods to obtain LST included the Radiative Transfer Equation method, the Single-channel Algorithm and the Split Window Algorithm (Sobrino et al. 2004; Jimenez-Munoz et al. 2014; Rozenstein et al. 2014). Studies showed that different land cover had different effects on heat radiation. Impervious surfaces released large amounts of heat while vegetation and water released less (Zhou and Dong 2021). There were many researches focused on the quantitative relationship between urban landscape pattern and LST. However, the remote sensing data used to obtain land cover classification, vegetation index, impervious surface were mainly medium resolution such as MODIS data (Chen et al. 2012a). The landscape index was applied to analyze the urban spatial form based on Class Level and Landscape Level (Wang et al. 2017).

Nan'an District was an important part of Chongqing with rapid economic development and high construction intensity. In this paper, we calculated the LST of Nan'an District based on the Radiation Transfer Equation method using the Landsat-8 data in 2021. Then we analyzed the LST of different land cover and carried out the correlation between land cover and LST by grid size of 100 m × 100 m using Geographical Conditions Monitoring Data. Landscape index was calculated to analyze the characteristics of LST under different landscape patterns such as patch density (PD), largest patch index (LPI), edge density (ED), Shannon's diversity index (SHDI) and aggregation index (AI).

9.2 Methodology

9.2.1 Study Area

Nan'an District was located in the west of Chongqing (106° 03′ 14″–106° 47′ 02″ E, 29° 27′ 02″–29° 37′ 02″ N) covering an area of 262 km^2 (Fig. 9.1). The Nan'an area was rich in heat and rainfall with an annual average temperature of 18.5 °C and rainfall of 1097.8 mm. We chose the Landsat-8 data with good weather conditions in August 2, 2021. According to previous studies, the tenth thermal infrared band of Landsat-8 had been widely applied to estimate the land surface temperature.

The preprocessing of Landsat-8 data included radiometric calibration, atmospheric correction and ortho-rectification (Fig. 9.2). The LST was obtained through the calculation of normalized difference vegetation index (NDVI), vegetation coverage (VC), emissivity and other parameters. The land cover from Geographical Conditions Monitoring Data consisted of 8 categories: planting vegetation, forest and grass, building(s), railway and road, structure, excavation surface, bare surface and water. The elevation for atmospheric correction was calculated from 1:50,000 DEM. The land cover data was rasterized to the grid size of 100 m × 100 m for analyzing the correlation with LST. Finally, the landscape index calculated from land cover data was applied to describe the characteristics of landscape spatial pattern. The

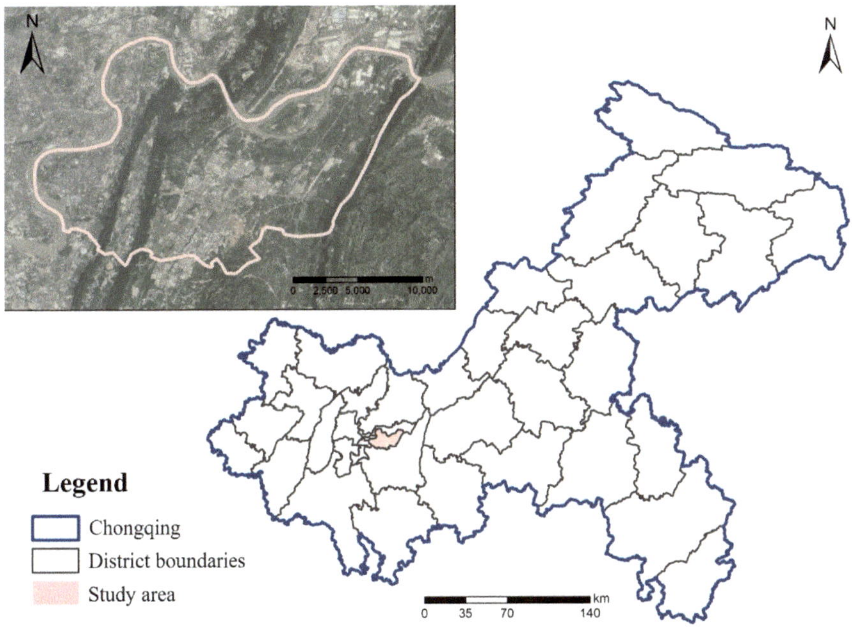

Fig. 9.1 Study area

coordinates of all the data were transformed into China Geodetic Coordinate System 2000.

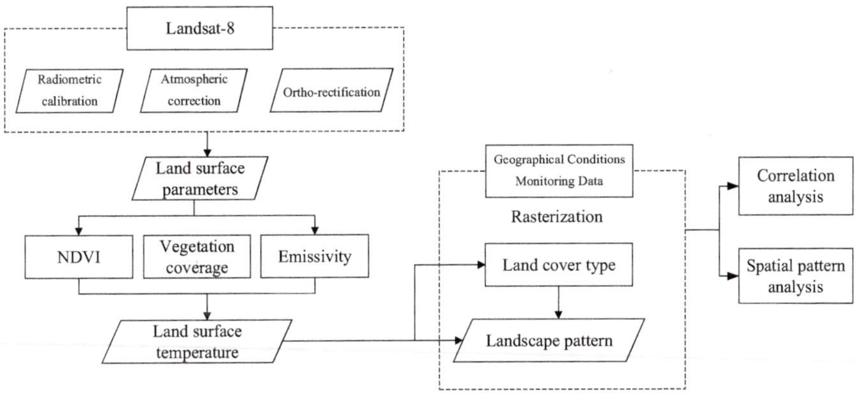

Fig. 9.2 Technical route of the paper

9.2.2 Estimation of LST

The software of ENVI was applied do the preprocessing of Landsat-8 data. The formula of LST by the Radiation Transfer Equation method was listed as follows:

$$T_s = K_1/ln(K_2/B(T_s) + 1) \tag{9.1}$$

$$B(T_s) = L_\lambda - L\uparrow - \tau(1-\varepsilon)L\downarrow/\tau\varepsilon \tag{9.2}$$

where, L_λ was thermal infrared radiance received by satellite sensor, T_s was LST, τ was atmospheric transmittance, ε was land surface emissivity, $L\uparrow$, $L\downarrow$ was ascending and descending radiance of atmosphere, respectively. K_1, K_2 was calibration constant represented 774.89 W m^{-2} sr^{-1} um^{-1}, 1321.08 K. $L\uparrow$, $L\downarrow$ and τ could be obtained from http://atmcorr.gsfc.nasa.gov/ with value of 3.30 W m^{-2} sr^{-1} um^{-1}, 5.12 W m^{-2} sr^{-1} um^{-1} and 0.62. The vegetation coverage (VC) was calculated from:

$$VC = [((NDVI - NDVI_s))/((NDVI_v - NDVI_s))] \tag{9.3}$$

where, the $NDVI_s$ was the NDVI of bare soil, the $NDVI_v$ was the NDVI with full vegetation coverage. According to previous studies (Qin et al. 2003), when NDVI was greater than 0.7, less than 0.05, VC could be taken as 1, 0. In order to calculate the land surface emissivity, the land cover was divided into three types: water, urban and natural surface. We got the emissivity of water as 0.995. The emissivity of natural surface and urban surface could be obtained as follows:

$$\varepsilon_{nature\_surface} = 0.9625 + 0.0614VC - 0.0461VC^2 \tag{9.4}$$

$$\varepsilon_{urban\_surface} = 0.9589 + 0.086VC - 0.0671VC^2 \tag{9.5}$$

From the estimation, we could get the average land surface temperature of Nan'an District (39.55 °C) which was very consistent with the actual temperature published by Chongqing Meteorological Bureau (http://www.tianqihoubao.com/lishi/chongqing/20210802.html) with average temperature of 40 °C. The LST was divided into 7 grades according to the Mean Standard Deviation method (Chen and Wang 2009). The result showed, the LST of Nan'an District was quietly high with maximum temperature of 57.63 °C and minimum temperature of 29.38 °C. The area of LST from 35.89 to 43.21 °C accounted for more than 71% of Nan'an District, and the area and the area from 37.72 to 39.95 °C accounted for 19.78% (Fig. 9.3).

The LST presented a low temperature zone in the peripheral area along the Yangtze River and in the striped area running through north and south. Because water had a significant cooling effect on thermal environment. Tongluo and Mingyue Mountain through north and south were mainly composed of forest and grass which also had a certain cooling effect. High LST areas were distributed in west and middle

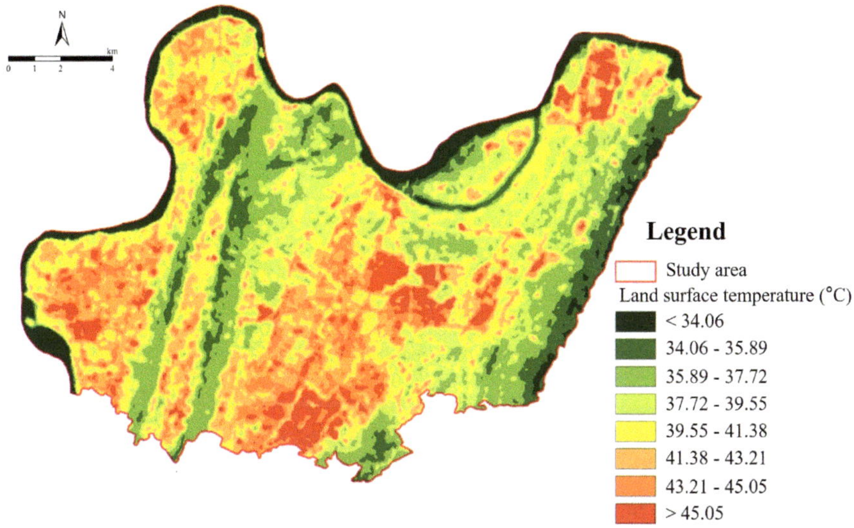

Fig. 9.3 Estimation of land surface temperature of Nan'an District

agglomerated in urban construction area such as building(s), railway and road, and structure.

9.3 Analysis Between Land Cover and LST

9.3.1 Characteristics of LST Under Different Land Cover

From the overlay of land cover data and LST data, it was concluded that LST range of building(s), railway and road, structure and excavation surface was from 39 to 45 °C (Fig. 9.4). LST range of planting vegetation, forest and grass, bare surface was from 35 to 42 °C. The water had the lowest LST range (29–35 °C). It was showed, the LST range of urban construction areas seemed significantly higher than the areas covered by water and vegetation. The cooling effect of water was very obvious which was 9 °C lower than building(s), railway and road, structure and excavation surface. Also, the area covered by planting vegetation, forest and grass a significant cooling effect that was 3 °C lower.

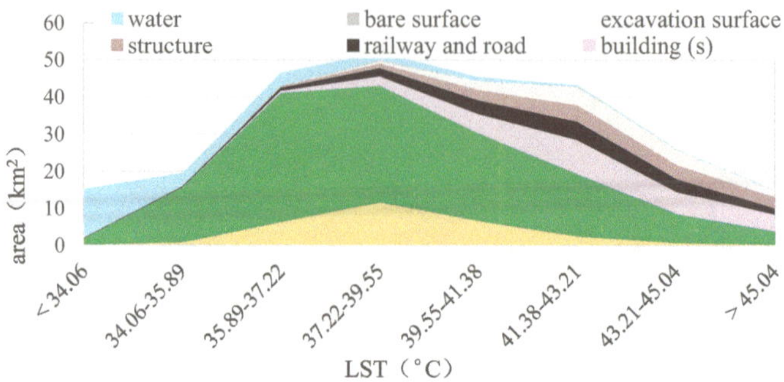

Fig. 9.4 Characteristics of LST under different land cover types

9.3.2 *Relationship Between LST and Land Cover*

The grid size of 50 m, 100 m, 150 m, and 200 m was constructed to analyze the relationship between LST and land cover, respectively. The correlation coefficient of planting vegetation, building(s) and structure increased as the grid size increased. However, the correlation coefficient of forest and grass, excavation surface, and water firstly increased then decreased with the increase of grid size (Table 9.1). Combined with the resolution of Landsat-8 data and the correlation between land cover and LST under different grid scales, a grid size of 100 m × 100 m was selected to conduct Pearson correlation analysis.

It was showed that water had a strong negative effect on LST with correlation coefficient of −0.7998 Forest and grass had a significant negative effect that absolute value was less than water (−0.5109). Building(s) had a positive correlation with LST (0.5752). The correlation between bare surface, planting vegetation and LST was not obvious and the absolute value of the correlation coefficient was less than 0.1. It was concluded that increasing the area of water, forest and grass could cool the ground surface efficiently while increasing the area of building(s), railways and roads could conversely warm the ground surface.

Table 9.1 Correlation between land cover types and LST at different grid scales

Grid scale (m)		50	100	150	200
Pearson correlation	Planting vegetation	−0.0380	−0.0567	−0.0889	−0.0940
	Forest and grass	−0.4511	−0.5109	−0.5169	−0.5095
	Building(s)	0.4332	0.5292	0.5752	0.5934
	Railway and road	0.3073	0.3331	0.3589	0.3962
	Structure	0.2077	0.3154	0.3837	0.4201
	Excavation surface	0.2813	0.3325	0.3379	0.3369
	Bare surface	−0.0442	−0.0024	−0.0125	−0.0479
	Water	−0.7574	−0.7998	−0.7822	−0.6960

9.4 Analysis Between Landscape and LST

In this paper, the landscape index of patch density (PD), largest patch index (LPI), edge density (ED), Shannon's diversity index (SHDI) and aggregation index (AI) was introduced to analyze the characteristics of landscape pattern (Chen et al. 2012b). Where, PD reflected the spatial pattern of the landscape which increased as the landscape heterogeneity became strong. LPI reflected the proportion of the largest patch area. ED reflected the landscape edge effect, the greater the edge density was, the higher the degree of patch complexity became. SHDI reflected the level of landscape heterogeneity and diversity, the value 0 represented only one species, a larger value indicated that each patch type was balanced in the landscape distribution. AI reflected the degree of agglomeration or extension trend of landscape that larger value represented higher landscape aggregation. The formulas of landscape index were as follows:

$$\mathrm{PD} = \frac{N}{A} \tag{9.6}$$

$$\mathrm{LPI} = \frac{max(a_1 \ldots\ldots a_n)}{A}(100) \tag{9.7}$$

$$\mathrm{ED} = \frac{\sum_{j=1}^{m} e_{ij}}{a_i} \tag{9.8}$$

$$\mathrm{SHDI} = -\sum_{i=1}^{m}[P_i ln(P_i)] \tag{9.9}$$

$$\mathrm{AI} = g_{ii}/g_{max} \tag{9.10}$$

where, N was the number of patches, A was total area of patches, a_i was area of the patch i, e_{ij} was the perimeter of patch j in patch type i, P_i was the area proportion of patch type i, g_{ii} was number of adjacent pixels in the same patch type, g_{max} was maximum possible number of adjacent pixels in the same patch type. The moving window algorithm in Fragstats software was applied to calculated the landscape index based on window size of 100 m × 100 m (Fig. 9.5).

The result showed there was a large number of land cover patches resulted in landscape fragmentation. The maximum value of PD reached 20/ha, and the average patch density was 4.86/ha mainly because of the urban construction of Nan'an District. The LPI seemed higher in water and forest areas and lower in urban areas such as building(s) and structure. The ED varied greatly in different land cover types with high value in forest and grass (288.17) and low value in water (93.42). It was because the shape boundaries of water seemed uncomplicated than other land cover types. The SHDI was characterized by a normal distribution with value range of [0–1.93]. The range of [0.6–1.2] accounted for 46.81% of the total area. Therefore, nearly half area of the Nan'an District represented balanced distribution of land cover patches.

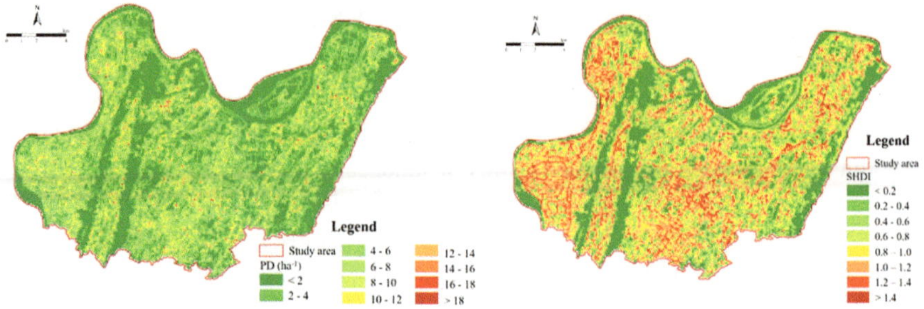

Fig. 9.5 Distribution of landscape index (left: PD, right: SHDI)

Table 9.2 Relationship between landscape index with LST

Landscape index	Pearson correlation coefficient with LST
PD	0.4207
ED	0.4961
SHDI	0.5753
AI	0.4535

Different land cover types had certain differences in AI with highest value in water (93.86) and lowest in railway and road (70.44). Because the water of the Yangtze River was widely distributed which caused the pattern index of AI generally high.

The relationship between landscape pattern and LST was calculated by the grid size of 100 m × 100 m (Table 9.2). The result showed that the landscape index of PD, ED and AI had a certain correlation with LST, but each correlation coefficient was less than 0.5. The SHDI had the highest correlation coefficient with LST with coefficients of 0.5764. Therefore, the improvement of landscape pattern could benefit to cooling the ground surface, but it was not as effective as directly changing the type of land cover.

9.5 Result and Conclusion

In this paper, the Radiative Transfer Equation method was introduced to estimate the land surface temperature of Nan'an District based on Landsat-8 data in 2021. Then Geographical Conditions Monitoring Data was applied to calculate the relationship between land cover and LST using a grid size of 100 m × 100 m. The landscape indexes obtained by land cover data such as patch density, largest patch index, edge density, Shannon's diversity index and aggregation index were applied to analyze the spatial pattern of landscape and the correlations with LST. There were some conclusions as follows:

(1) The LST in Nan'an District was relatively high as a whole, with average temperature of 39.55 °C in August 2021. High LST areas distributed in urban construction areas such as building(s), railway and road, structure, while low LST areas were mainly located in the peripheral area along the Yangtze River and in the striped area running through north and south mainly covered by forest and grass.
(2) Among all the land cover types, the LST areas of building(s), railway and road, structure were 9 °C higher than water and 3 °C higher than planting vegetation and forest and grass. It was concluded that water had significant cooling effect and ground surface covered by vegetation also had a certain cooling effect.
(3) It was showed that water had a strong negative correlation with LST while building(s) had a positive correlation. Therefore, increasing the area of water, forest and grass could cool the ground surface efficiently and increasing the area of building(s) and structure could warm the ground surface.
(4) There were large patches in Nan'an District resulted in high fragmentation. The ED of forest and grass was larger than other land cover type. The AI of water area represented quite high with value of 93.86. All the landscape indexes had a certain correlation with LST, so improving the landscape pattern could benefit to cooling the ground surface.

However, there were some shortcomings in this research. Firstly, the Geographical Conditions Monitoring Data was obtained in June 2021 and Landsat-8 data in August 2021 which could affect the relationship analysis. Secondly, the big difference of resolution of each data may have a certain impact on the research results.

Acknowledgements This work is mainly supported by Research on key technologies of remote sensing monitoring for ecological restoration of open-pit mine based on LiDAR (Grant No. CSTB2022NSCQ-MSX1484) and the Science and Technology Research Program of Chongqing Municipal Education Commission (Grant No. KJQN202103410).

References

Chen S, Wang T (2009) Comparison analyses of equal interval method and mean standard deviation method used to delimitate urban heat island. J Geo-Inf Sci 11(02):145–150

Chen A, Sun R, Chen L (2012a) Applicability of traditional landscape metrics in evaluating urban heat island effect. Chin J Appl Ecol 23(08):2077–2086

Chen A, Sun R, Chen L (2012b) Studies on urban heat island from a landscape pattern view: a review. Acta Ecol Sin 32(14):4553–4565

Cui L, Li G, Jian D (2018) Heat island effect and its relationship with land use in Chengdu City. Chin J Ecol 37(05):1518–1526

Jimenez-Munoz JC, Sobrino JA, Skokovic D et al (2014) Land surface temperature retrieval methods from Landsat-8 thermal infrared sensor data. IEEE Geosci Remote Sens Lett 11(10):1840–1843

Qin Z, Li W, Zhang M et al (2003) Estimating of the essential atmospheric parameters of mono—window algorithm for land surface temperature retrieval from Landsat TM6. Remote Sens Land Resour 02:37–43

Qiu X, Gu L, Zeng Y et al (2008) Study on urban heat island effect of Nanjing. Clim Environ Res 13(06):807–814

Rozenstein O, Qin Z, Derimian Y et al (2014) Derivation of land surface temperature for Landsat-8 TIRS using a split window algorithm. Sensors 14(4):5768–5780

Sobrino JA, Jiménez-Muñoz JC, Paolini L (2004) Land surface temperature retrieval from Landsat TM 5. Remote Sens Environ 90(4):434–440

Wang X, Yu D, Cao Q et al (2017) Quantitative relationship between urban landscape pattern and surface temperature. J Beijing Norm Univ (Nat Sci) 53(03):329–336

Zhao Z (2008) The analyse on urban heat-island effect factors and the countermeasures on the issue. Environ Monit China 24(06):77–79

Zhou H, Dong F (2021) Research on response of land use and surface temperature in Nan'an District of Chongqing City. J Anhui Agric Sci 49(15):79–84

Chapter 10
Bioclimatic Strategies and User Comfort in Homes in the Simon Bolivar District—Cerro de Pasco 2022

Esenarro Doris, Vanessa Raymundo, Dayana Chipana, Pamela Gavidia, and Pablo Cobeñas

Abstract The objective of this research is to propose bioclimatic strategies that allow the comfort of users, the district has evidence of air pollution generated by mining companies in the last 40 years that causes the inhabitants to develop toxic diseases, as well as the lack of living spaces that provide comfort to its inhabitants. As a methodology, bioclimatic design strategies were applied that respond to the climatology and topography of the place, also a historical analysis of its construction systems and the levels of contamination by heavy metals in the study site was developed, the processing of the information was supported by digital tools (AutoCAD, Revit, Excel). The results obtained in the proposal are the capture of solar radiation supported by the thick mud and stone walls that works as thermal insulation, as well as the thermal sheet floors in the rest areas to generate thermal insulation, which is a very thin and flexible laminated electric heater that allows distributing heat throughout the surface, the uniform heat rises to a maximum of 28 °C, also the incorporation of flora that allows minimizing polluting metals caused by mining found in an area. In conclusion, an adequate design with ecological materials allows to provide comfort to the users, making use of clean energies, and taking care of the environment.

Keywords Quality of life · Integration of public spaces · Bioclimatic strategies

E. Doris (✉) · V. Raymundo · D. Chipana · P. Gavidia · P. Cobeñas
University Ricardo Palma, Lima, Peru
e-mail: doris.esenarro@urp.edu.pe

E. Doris
Specialized Institute for Research on Ecosystems and Natural Resources - INERN, Lima, Peru

C. Yuan et al. (eds.), *Proceedings of 4th International Conference on Resources and Environmental Research—ICRER 2022*, Environmental Science and Engineering,
https://doi.org/10.1007/978-3-031-31808-5_10

10.1 Introduction

From the beginning of the settlement of the city of Cerro de Pasco, its first settlers were linked to mining activities, which is why they formed groups dedicated to mineral extraction; mining is an economic activity on which the national government and foreign companies have bet heavily, due to the great mining potential of Cerro de Pasco, which was promoted as a source of important economic resources for national development, however, it was not to the benefit of its inhabitants. Socio-environmental conflicts associated with large-scale mining constitute one of the fundamental issues in the contemporary mining debate, and numerous studies show the growing conflicts associated with large mega-mining projects, especially in Latin America (Alier 2001). Conflicts between mining companies and affected communities have been mainly linked to environmental impacts, land disputes, human rights violations (Sánchez-Vázquez et al. 2016a), as well as a lack of compliance with Corporate Social Responsibility Policies. The tensions surrounding mining activities are concentrated in what the Ombudsman's Office calls "socio-environmental conflicts", which involve confrontations between the local population and the companies operating in the territories they occupy, including disputes over land, water, means of transportation, contamination, and vulnerability (García 2006). Recently, several theses have sought to understand the weight of mining production and exports in the country's economic growth (Hurtado et al. 2012), while a number of research studies seek to problematize this claim using more refined methods (Sánchez-Vázquez et al. 2016b), exploring the causal relationship between mining and user welfare.

Since 1960, multiple international treaties have been signed promoting the establishment of the concept of sustainable development (Glave and Kuramoto), understood as "that which guarantees the needs of the present without compromising the possibilities of future generations to meet their own needs" (Díaz Palacios et al. 2016). The precariousness of constructions and buildings, as well as the physical location of populations and urban infrastructures in risk areas reflects the high level of vulnerability to a natural phenomenon (Díaz et al. 2016). It is necessary to promote effective control and sanction practices in coordination with internal control, the comptroller's office and the Public Prosecutor's Office when recommendations and guidelines on land use, planning, zoning, urban parameters and others are not taken into account (Bedoya 2003). Since this problem has consequences on the comfort of users, represented from their land use planning, housing planning, self-built housing, lack of living spaces, among others (Bedoya 2011).

Figure 10.1a shows that the city accentuates its growth in a disorderly manner, without urban planning (Gobierno Regional de Pasco, Programa Regional de Población 2012–2017). Figure 10.1b shows the beginning of a differentiation of population strata where the elite population located their homes in the central part of the city (Ministerio de Transportes y Comunicaciones 2005), Fig. 10.1c, d shows the urban growth of the city maintaining the same concept with which it had always been formed (Municipalidad Provincial de Pasco 2006).

Fig. 10.1 Typology of traditional housing in the Cerro de Pasco region

The biggest environmental problem found in the Cerro de Pasco region is the mineral deposits or tailings resulting from processing. Likewise, there is inadequate management of all urban and industrial solid waste (Federico Helfgott Cerro de Pasco 2012).

As a consequence of the expansion of mining and industrial activities, there is a disorderly and dispersed growth of urban and peripheral population centers that makes it difficult to provide public services (INGEMMET 2011). There is urban disarticulation between the three districts that make up the city, Simón Bolívar, Chaupimarca and Yanacancha, and the use of space and natural resources has generated a conflict of interests between the mining company, the Peruvian State and the inhabitants of the city and the periphery (Volcan Compañía Minera S.A.A. Memoria Anual 2005). The constant expansion of the mine generates a reduction in urban space and deficiencies in basic infrastructure, sanitation and electrification (Volcan Compañía Minera S.A.A. Memoria Anual 2005). The investment in housing infrastructure by the company for its workers and their families generates a differentiation with the inhabitants who do not have a direct labor relationship, which polarizes the city (Programa de Adecuación y Manejo Ambiental 1997). There is a lack of housing and equipment replacement programs for all urban areas near the operations being demolished (Nardin Daniel 1993).

Most of the territory of the city of Cerro de Pasco is part of the informal population, so they are not considered part of the formal city (Carhuaricra Meza 1996). Their precarious dwellings are located in the areas with the most rugged and rough topography, as they cannot be located within the urban center of the city due to the

Fig. 10.2 Typology of housing in the Chaupimarca sector in the Cerro de Pasco region

lack of available land. The constructions have been built following the contour lines of the hillsides, which means that they have only one front, since the back is attached to the hill.

Figure 10.2a shows the Chaupimarca sector, the oldest district in the city and the most sensitive to the mine's operational expansions. Its activity is mostly a mix of housing and commercial use. Figure 10.2b shows the accumulation of solid waste on the hills that are covered with mining-industrial waste very close to the houses, which is a polluting factor for all the people living in the area.

10.1.1 Literature Review

Bioclimatic Strategies. It is designed to take advantage of climatic and environmental conditions to achieve a situation of indoor thermal comfort (Oroya 2011). It plays exclusively with the design and architectural elements, without the need to use mechanical systems or energy inputs, it allows to obtain thermal comfort conditions in a house taking advantage of natural resources (Pajuelo 2005). In this way, it is possible to air-condition the house with a much higher energy efficiency.

Malecón Ribereño. We are faced with the need to extend the standards that determined the quality of urban spaces generated by the deployment of the industrial model; we need to extend them to the rest of the dimensions demanded by the citizen (Toni 2009). The indefinite growth of consumption cannot solve the environmental deficiencies of our surroundings (nor the destruction of the global environment), nor the alienation of a citizen who does not participate in the construction of his environment. The quality of urban life is the concretion of the Quality of Life on the urban space, being able to be considered as a social construct formed by three basic dimensions: Environmental Quality, Well-being, Identity (https://www.lhistoria.com/peru/cerro-de-pasco).

Contamination. Pollution is a determining factor in the deterioration of the quality of life (health, clean air, clean water, recreation, nature conservation, etc.). All human activities and natural phenomena are accompanied by emissions of gases, vapors,

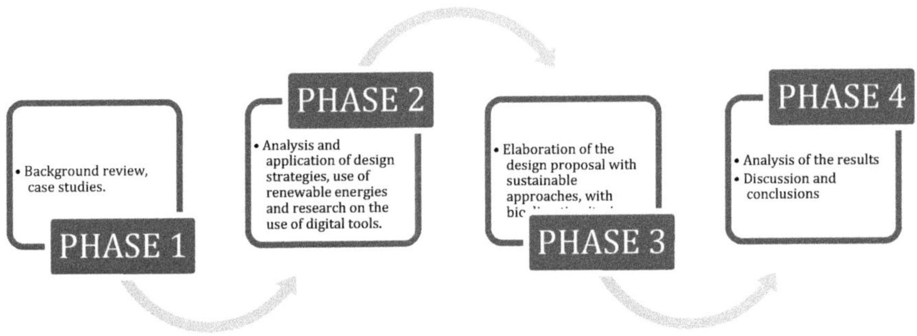

Fig. 10.3 Phase study

dusts, and particles. Pollution is always a negative alteration of the environment that mainly affects the basic natural resources: air, soil and water (https://inspiracion.ciep.edu.pe/personajes-de-peru/pasco/).

Mining. Mining continues to be necessary for the development of society and influential in the economy of Latin American and Caribbean countries, but public policies are not consistent in improving socio-environmental management (Tromme). In general, from the communities' point of view, mining is not currently seen as a development factor in their territories; on the contrary, they consider it to be an activity that pollutes water resources, generates atmospheric emissions and produces negative health effects, with a very low impact on economic development (Sennett 1975).

10.2 Methodology

10.2.1 Methodological Scheme

Figure 10.3 shows the analysis of the field survey for the preparation of the report with a compilation of information from matrices to make a proposal.

10.2.2 Location of the Study Site

Cerro de Pasco, located in the treeless Peruvian highlands, is one of the cities located at the highest altitude on the planet, at 10° south latitude and at an altitude of 4380 m above sea level. The city of Cerro de Pasco is located in the Province of Pasco and in the Pasco Region. It is located in what corresponds to the natural region of Puna, which is characterized by scarce minor vegetation and the non-existence of trees.

Fig. 10.4 Map showing the location of the study site in the Cerro de Pasco region

The city extends over a plain surrounded by hills and extensive pampas of ichu or thatch (León 2006).

Figure 10.4 shows the geographic location and delimitation of the study area, which is located in the Bombón Plateau, an extensive plain that extends into the department of Junín.

10.2.3 Climate

Its climate is quite cold. The average maximum temperature is 12.4 °C while the average minimum temperature is 0.6 °C. The pluvial precipitations are important between the months of October and April being the average of 1182.7 mm. With respect to the winds, these are of moderate intensity that go in a northeasterly direction, especially in the afternoons, after 16:00 h (Foto Archivo. Revista Digital de Arquitectura).

Humidity and Temperature. Relative humidity is inversely proportional to temperature, the higher the humidity the lower the temperature and the lower the humidity the higher the temperature, it was analyzed that the maximum humidity occurs in the early morning while the minimum temperature is experienced (Rueda 2001).

Figure 10.5a shows that relative humidity is inversely proportional to temperature, the higher the humidity the lower the temperature and the lower the humidity the higher the temperature, and in Fig. 10.5b it was analyzed that the maximum humidity occurs in the early morning hours while the minimum temperature is experienced (Rueda 2001).

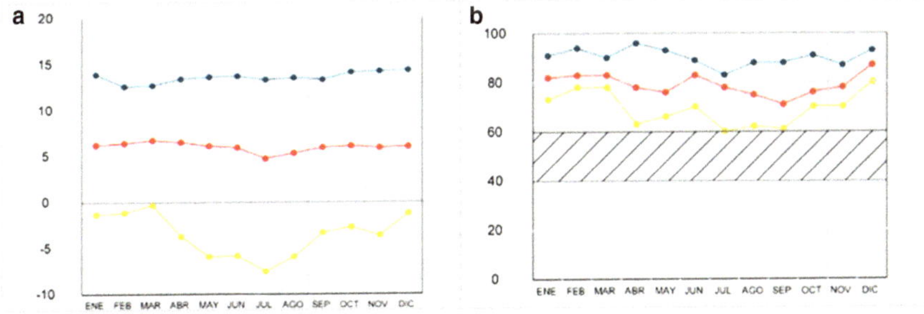

Fig. 10.5 **a** Temperature, **b** humidity in the Cerro de Pasco region

Precipitation and Winds. Rainfall is observed in the summer months of December, January and February, as well as in March and in the preceding months of October and November, a total of 6 months with rainfall in the "rainy" range; in the other months of the year, rainfall is also observed, but of lesser intensity since it is located in the low rainfall range.

Figure 10.6a shows the rainfall in the summer months of December, January and February, as well as in March and in the preceding months of October and November, a total of 6 months with rainfall in the "rainy" range, In the other months of the year there are also rains but of less intensity since they are located in the range of few rains or showers, and in Fig. 10.6b we can observe the summary of the wind measurements of the 7, 13 and 19 h, determining the predominant winds for the design. In the case of the 7 h the prevailing winds come from the north in 8 months of the year, in the remaining four there is no wind (calm). In the 13- and 19-h wind measurements, the wind is predominantly from the north all year round (MINSA 2015; Proderm, Gobierno Regional Pasco 2024).

Table 10.1 shows the wind measurements for Cerro de Pasco, showing that the most frequent winds occur between 7:00 a.m., which is of low intensity and comes from the north, and also at 1:00 p.m. there is medium intensity, these winds come

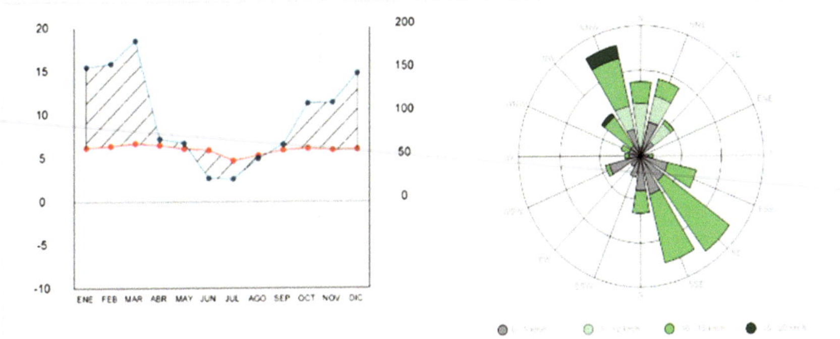

Fig. 10.6 Natural physical analysis—precipitation and wind measurement

from the north, the speed fluctuates between 3.5 and 6.5 m per second and finally the winds at 7:00 p.m. are of less intensity, these come from the north, the speed fluctuates between 2.0 and 2.5 m per second (Proderm, Gobierno Regional Pasco, 2024).

Solar Geometry. The solar geometry analysis has the summer months December to March with few hours of sunshine, between 3 and 4 h of sunshine per day, because it has a cloudy sky, the rest of the year the sky of the city is in the range of semi cloudy, the months of June, July and August are higher solar radiation, the solar analysis is shown according to the schedules made in the Revit programs (Nacional and de Planeamiento Estratégico—CEPLAN 2011).

In Fig. 10.7a, the solar geometry analysis shows the summer months from December to March with few hours of sunshine, between 3 and 4 h of sunshine per day, due to a cloudy sky, the rest of the year the sky of the city is in the range of semi cloudy, the months of June, July and August are of higher solar radiation, and in Fig. 10.7b, shows the solar analysis according to the schedules made in the Revit programs (Nacional and de Planeamiento Estratégico—CEPLAN 2011).

Comfort Table—Givoni. In Cerro de Pasco, the summer months, from December to March, have few hours of sunshine, between 3 and 4 h of sunshine per day, due to cloudy skies; the rest of the year the city's skies are in the semi-cloudy range.

Table 10.1 Wind measurements of Cerro de Pasco (Nacional and de Planeamiento Estratégico—CEPLAN 2011)

	Hour	Jan	Feb	Mar	Apr	May	Jun	Jul	Ago	Sep	Oct	Nov	Dec
Winds (m/s)	07:00 m/s	13.9	12.6	12.7	13.4	13.6	13.7	13.3	13.5	13.3	14.1	14.2	14.3
	13:00 m/s	−1.4	−1.2	−0.4	−3.8	−6.0	−5.9	−7.6	−6.0	−3.4	−2.8	−3.7	−1.3
	19:00 m/s	9173	9478	9078	9663	9366	8970	8360	8862	8861	9170	8770	9380

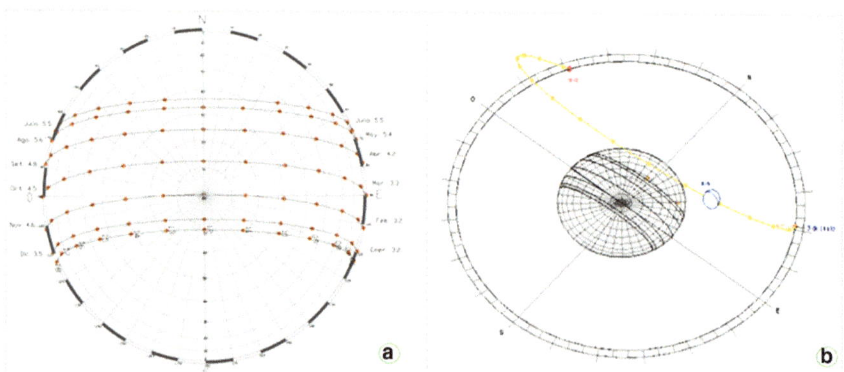

Fig. 10.7 Natural physical analysis image—solar geometry of the Cerro de Pasco region

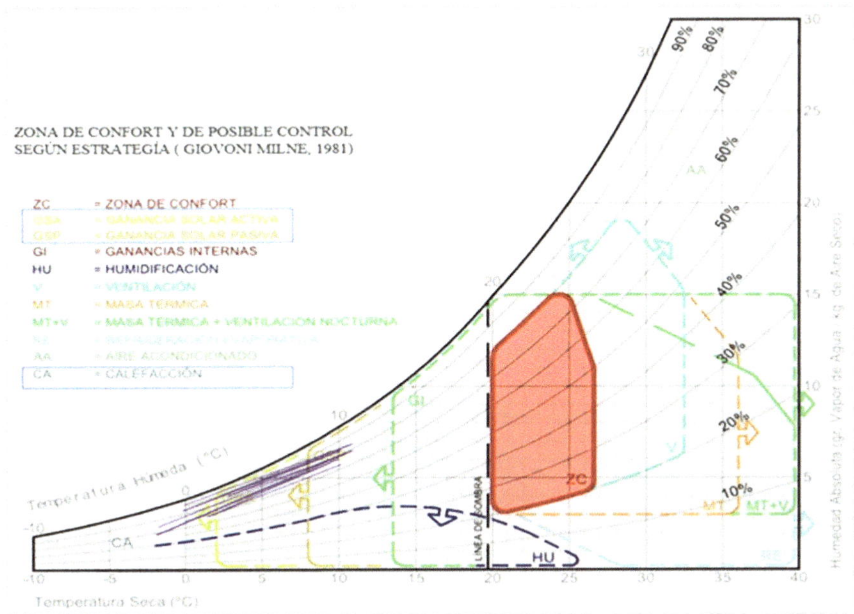

Fig. 10.8 Natural physical analysis—comfort graph—Givoni of the Cerro de Pasco region

Figure 10.8 shows the analysis of the Confort-Givoni chart for the months from January to December, showing the months with the highest active and passive solar gain (Mundial 2012).

10.2.4 *Flora and Fauna of Cerro de Pasco*

They vary according to the different natural regions of its territory, from the punas to the lowland jungle, passing through all the intermediate regions. This diversity of environments creates a great variety of ecosystems that make the department an important natural reserve, which must be protected to avoid its depredation (Modelo para orientar la gestión hacia la Conservación y el Desarrollo Sostenible de una Región con Reserva Comunal 2016).

Flora and Fauna. At around 4,000 m, the quiñual or queñual trees (A) are remnants of ancient high-altitude forests. Among the grasses, ichu (Stipa festuca) (B) and chiligua (Calamagrostis) (C) predominate. Above 5,200 m, in addition to some mosses and lichens (D), the yareta (phanerogam) grows. In the jungle, the vegetation is tropical, with a great variety of trees such as moena (E), walnut (F), mahogany or aguano (G), tornillo, cedar, capaiba and ishpingo (H). In the jungle brow there are tropical cloud forests with several species of trees and a great variety of epiphytes and orchids (more than 500). The trees, covered with moss and lichens,

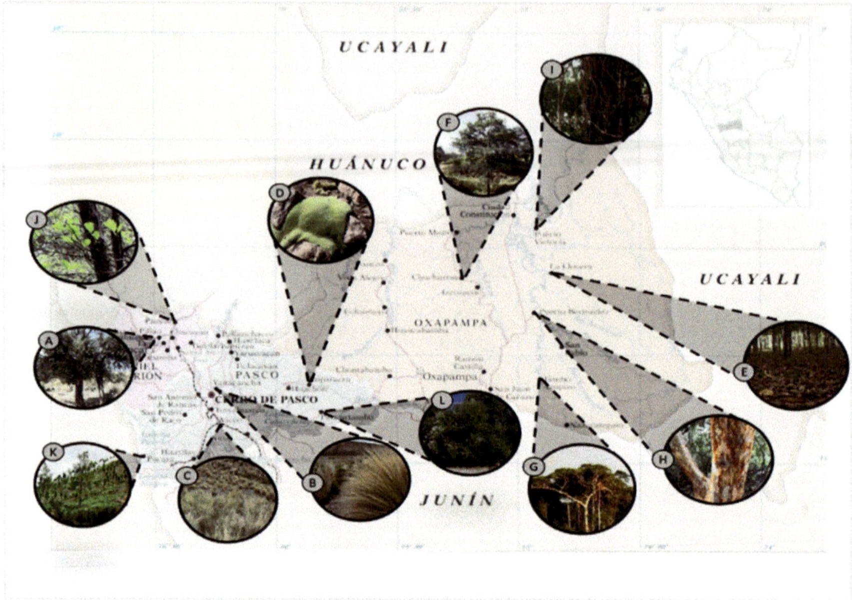

Fig. 10.9 Image of the diversity of flora species in the Cerro de Pasco region

are so numerous that their canopies intertwine. There is also a great variety of lianas (I), such as the tamushi or tamshi (Carludovica sp.), which climb all kinds of trunks and branches in their natural search for sunlight. In addition to the quiñual, the sierra is home to alder (Lambras sp.) (J), eucalyptus (Eucalyptus globulos) (K), cypress (Cyperus sp.), pine (Pinus radiata), kolle (in danger of extinction, so its logging has been banned) and willow (Salix chilensis) (L) (Rodriguez Larrain 2019).

Figure 10.9 shows the diversity of flora species according to the natural regions of its territory. In the highlands we find the queñual, ichu, grasses, alders, eucalyptus, and willows. While in the jungle we find the moena, walnut, mahogany, ishpingo and llana, among others (Rodriguez Larrain 2019).

10.2.5 Heavy Metal Mitigating Flora

Figure 10.10 shows the diversity of flora species that allow minimizing the contaminating metals caused by mining in an area, there are some species such as (a) Atriplex halimus (salado), (b) Hyparrhenia sinaica (cerrillo), (b) Hyparrhenia sinaica (cerrillo), (c) Lygeum spartum (albardín), (d) Piptatherum miliaceum (mijera), (e) Salsola oppositifolia (salao borde), (f) Suaeda vera (sosa), etc.

Fig. 10.10 Image of the diversity of heavy metal mitigating flora species

Fauna. In the highlands, the fauna is not as numerous and varied as in the Amazon. The Andean gray deer (Cdocoileus peruvianus) (A), the viscacha (Lagidium peruvianum) (B), the condor (Vultur gryphus) (C), the taruca (Hippocamelus antisesis) (D), the wild guinea pig (Cavia cobaya) (E) and others; There are also camelids such as llamas (Lama glama) (F) and vicuñas (Vicugna vicugna); other highland species include the wild cat, the skunk (G), the gallinazo and various birds of the partridge family.

In the Pasqueña jungle, animal diversity is much greater. There is a great variety of mammals, reptiles, birds and insects, such as the following: sachavaca (H), sajino (I), huangana or wild pig (J), deer, otorongo (K) or American tiger, margay, ronsoco or capybara, añuje or agouti, armadillo, peccary, anteater, squirrel, sloth (L), wild rabbits and a great variety of monkeys. The rivers are the habitat of beautiful mammals such as the manatee or sea cow and the otter or river otter (Rodriguez Larrain 2019), as well as a great variety of lizards and snakes (boa anaconda, afaninga and shushupe) (M).

Figure 10.11 shows the diversity of fauna species according to the natural regions of the territory. In the highlands, we find the Andean gray deer, viscacha, condor, taruca and wild guinea pig. In the jungle, we find the sachavaca, the sajino, the huangana, and the otorongo, among other species (Rodriguez Larrain 2019).

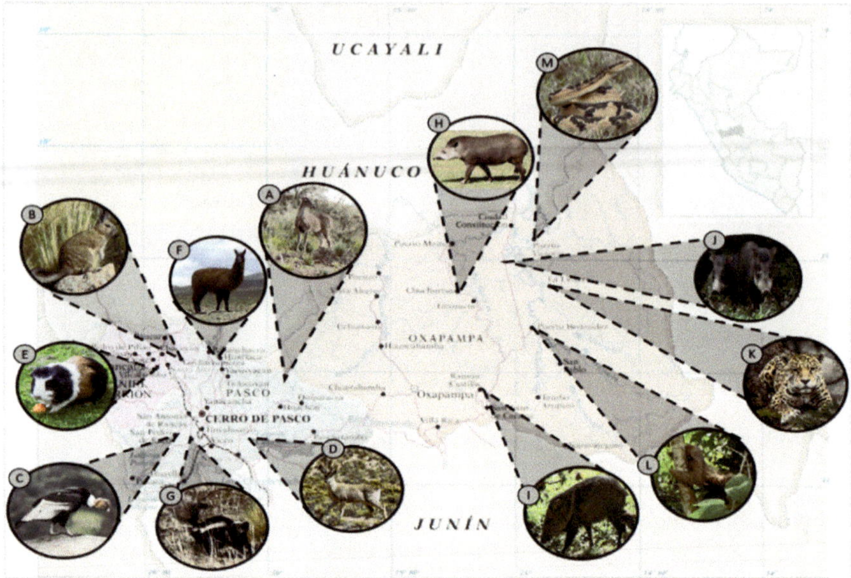

Fig. 10.11 Image of the diversity of fauna species in the Cerro de Pasco region

10.3 Results

10.3.1 Project Location

It is located in the Bombón Plateau, an extensive plain that extends to the department of Junín, with coordinates at latitude: S 10° 40′ 2.93″ and longitude: W 76° 15′ 24.05″ (Ochoa 2021).

Figure 10.12 shows the geographic location and delimitation of the study area, which is located in the Bombón Plateau, an extensive plain that extends to the department of Junín, with coordinates at latitude: S 10° 40′ 2.93″ and longitude: W 76° 15′ 24.05″ (Ochoa 2021).

10.3.2 Volumetric Proposal

Figure 10.13 shows an elevation of the house built with bioclimatic strategies that will be applied in the proposal of the house and the different materials that will be used to minimize the environmental impact that occurs in the region of Cerro de Pasco.

Fig. 10.12 Map of the study site, district of San Cristóbal-Cerro de Pasco

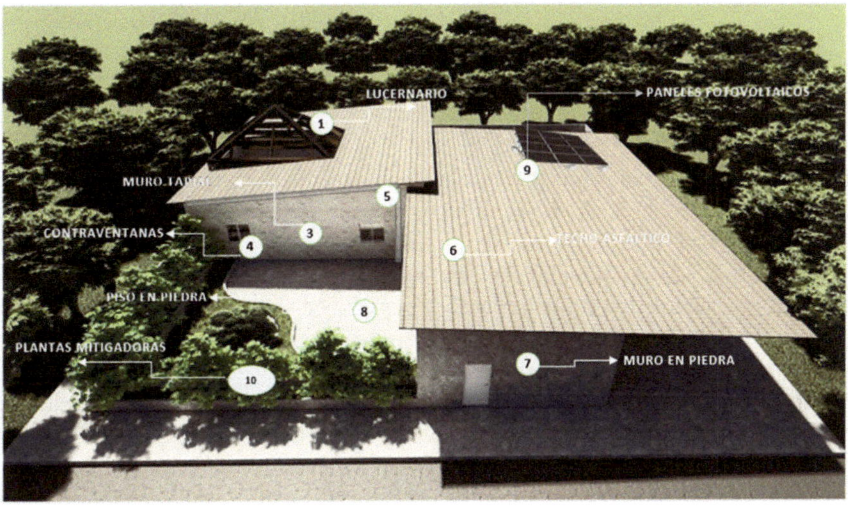

Fig. 10.13 Volumetric proposal of the 3D housing proposal

10.3.3 Construction System

Figure 10.14 shows the cut of the interior spaces with strategies of the materiality that will be used to generate comfort in the house during the different seasons of the year. Tapial will be used as a construction technique to take advantage of the thermal insulation of this material in internal environments to generate comfort and contrast

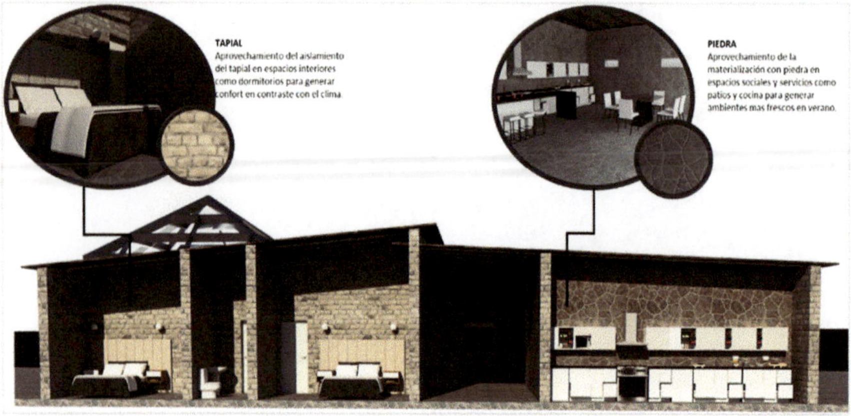

Fig. 10.14 Cutting of the interior spaces

the climate. Another material to be used is stone on the floors of the interior spaces to achieve cooler environments in summer.

10.3.4 Solar Path Analysis

In Cerro de Pasco, the summer months, from December to March, have few hours of sunshine, between 3 and 4 h of sunshine per day, due to cloudy skies; the rest of the year the city's skies are in the semi-cloudy range.

Figure 10.15 shows the solar projection during the months of highest temperature and sunshine hours: (a) Equinox: March/September, 8 a.m.; (b) Equinox: March/September, 3 p.m.; (c) Summer: January, 9 h; (d) Summer: January, 15 h.

Figure 10.16a shows how the central courtyard is illuminated through the skylight, which allows the sun's rays to enter and increase the comfort of the users inside the house, as they do not depend on artificial lighting during daylight hours. Natural light is beneficial for the human being and collaborates in the reduction of the energy saving effect. Figure 10.16b shows the entrance of lighting that allows having the kitchen-dining room space with good lighting and better ventilation and cooling inside the house, taking advantage of the outside wind to create natural ventilation, thanks to an exchange of air between the outside and inside of it and giving a good comfort to the user who is inside the space.

Figure 10.17a shows the incidence of the sun during the day and the entry of the winds for better ventilation, and to have a better heat input inside the house that is used for good comfort in each of the spaces and capture the radiation of the sun to the north and keep the heat through the thick walls of mud and stone, and store the heat produced during the day for the night that are hours that lower the temperature and allow to have the interiors of the rooms warmer. Figure 10.17b shows

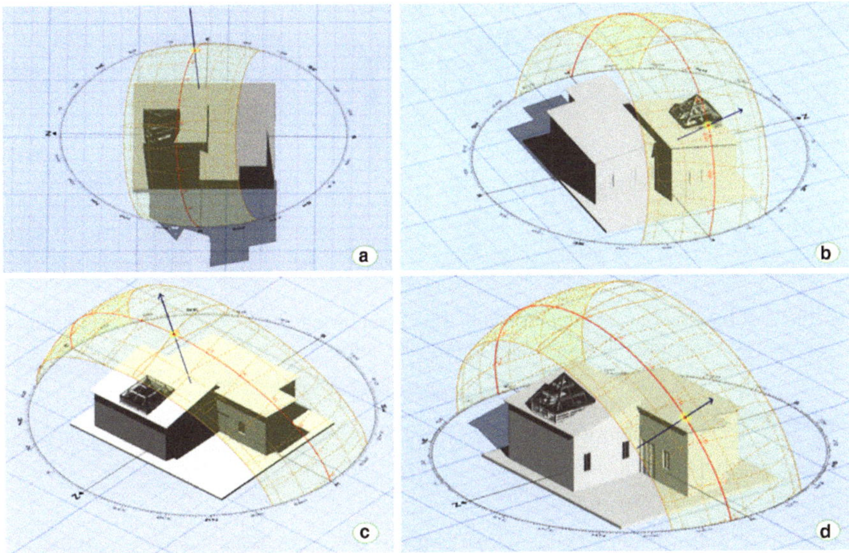

Fig. 10.15 Sunpath in the WebGL application

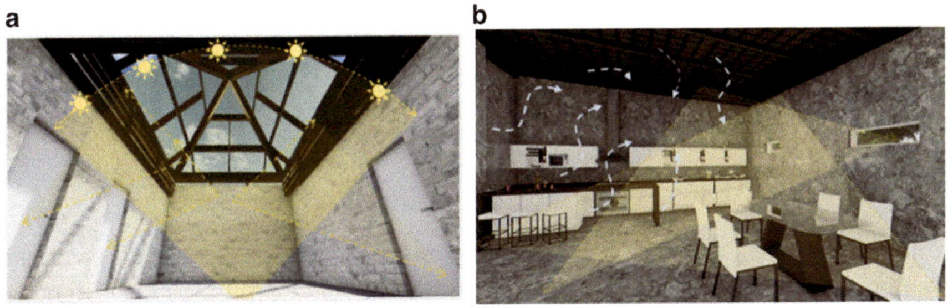

Fig. 10.16 **a** Interior view of skylight in patio of prototype house. **b** Interior view of kitchen-dining room

the materialization of the thermoplastic floors in the rest areas to generate thermal insulation, which is a very thin and flexible laminated electric heater that allows heat to be distributed over the entire surface, the uniform heat rises to a maximum of 28 °C (following radiant heating regulations), thus avoiding expansion gradients in the cold and hot areas and keeping the space more comfortable during the night due to the cold that occurs in Cerro de Pasco.

Table 10.2 shows the average thermal energy consumption for the floor, as a strategy to be applied for spaces that do not have much sun radiation during the day, and the thermo foil will allow during the night hours to keep the spaces more comfortable for the users that are inside these spaces.

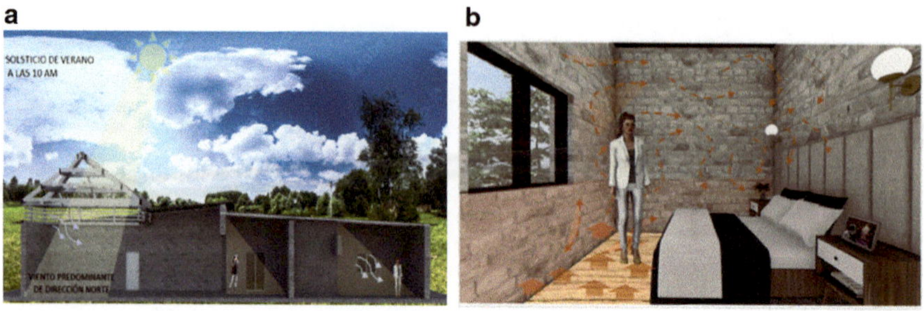

Fig. 10.17 **a** Interior view of the house with the sun and wind inlets, **b** Interior view of bedroom with prototype of thermo foil floor strategy inside the house

Table 10.2 Wind measurements of Cerro de Pasco (Proderm, Gobierno Regional Pasco 2024)

Average consumption of thermal floor				
	Termopiso			
Environment[a] (m)	Power (W)	Consumption (W)	Consumption (W)	Btus
8	1200	423	480	4.320
12	2400	648	720	6.480
18	3600	972	1080	9.720
20	3600	1,080	1200	10.800

[a] Height = 2.70 m—thermostat regulated in 18 °C—average external temperature 5 °C

10.4 Discussion

Approximately half of the households in Sonora have low or very low levels of access; the services with the lowest levels of access are thermal comfort, efficient lighting, and information and entertainment; this situation affects human rights, which is magnified by the confinement in the homes (Ochoa 2021), as well as the improvement of comfort and safety of the habitat through the study and development of techniques applied to the construction elements of rural housing. For technology transfer, the training of technical promoters in safe and healthy housing is proposed, as well as the generation of local enterprises around the implementation of the proposed solutions (Wieser et al. 2021), the proposal seeks thermal comfort and energy savings, in the face of the impacts of environmental pollution that occurs in Cerro de Pasco due to mining, taking advantage of, sustainability strategies to be applied within the housing that will provide ecosystem services, better thermal comfort, efficient lighting for user comfort and cultural exchange, which promotes preservation and environmental awareness, so the proposal considers design criteria and construction elements with clean, safe, healthy and sustainable technology where the environmental impact is minimized.

10.5 Conclusions

A prototype of a single-family house was developed considering the low environmental impact by being totally sustainable, through the use of design strategies such as the use of new alternative energies to obtain the necessary resources to provide comfort to the people who live in the place and to those who visit it.

The use of bioclimatic materials such as the use of partition walls, which is a good economic and ecological thermal insulator, the proper orientation, and the use of integral construction systems to positively influence the construction costs of the house, optimizing it.

To have a good comfort inside the house it is necessary to have a good natural ventilation, for this it is necessary to measure variables such as air temperature, radiation temperature, relative humidity of the air and air speed, in this way we can have spaces that allow to create a pleasant feeling for the users.

It was determined that it is possible to use passive systems such as: direct solar gain (passive lighting), greenhouse gain, high thermal inertia roofs, solar incidence control and cross ventilation, to achieve a coherent response to the context and the optimal development of comfort in the environments.

It was decided to include vegetation in the courtyard to improve air quality and create a barrier against environmental pollution.

References

Alier M (2001), Gudynas (2009), Bebbington et al. (2010)

Bedoya C (2003) El concreto reciclado con escombros como generador de hábitats urbanos sostenibles La ciudad como ecosistema semi-cerrado, una utopía cultural. Tesis de Maestría en Hábitat. Universidad Nacional de Colombia. Medellín. Recuperado de

Bedoya C (2011) Viviendas de interés social y prioritario sostenibles en Colombia -VISS y VI PS. Revista internacional de sostenibilidad, tecnología y humanismo 6:27–36

Carhuaricra Meza E (1996) Pasco: Realidad de acero, posibilidad de oro, Cerro de Pasco. Birz Editores

Centro Nacional de Planeamiento Estratégico—CEPLAN (2011) Plan Bicentenario. El Perú hacia el siglo XXI. (Perú), 265 pp. Disponible desde Internet en https://observatorioplanificacion.cepal.org/sites/default/files/plan/files/planbicentenarioceplanindex.pdf

Díaz M, Melgar D, Tapia B, Vallejo P (2016) Hacia un análisis de la gestión de ecoeficiencia minera: un estudio de seis empresas mineras en el Perú (tesis de maestría). Pontificia Universidad Católica del Perú, Lima, Perú

Díaz Palacios J, Arana Cardó M, Torres Guevara J, Patrucco S (2016) Historia ambiental del Perú. Siglos XVIII y XIX. Minam, Lima

Federico Helfgott Cerro de Pasco (2012) Historia y Espacio Urbano Lima, Perú

Foto Archivo. Revista Digital de Arquitectura

García (2006), Cadenas y Loayza (2019), Castillo & Dueñas (2019), Luna (2019)

Gobierno Regional de Pasco, Programa Regional de Población 2012–2017 Pasco—Perú

Glave MA, Juana Kuramoto, Grupo Análisis para el Desarrollo-GRADE Minería, Minerales y Desarrollo Sustentable en Perú, Capítulo 8. Lima, Perú

Historia (05 de mayo de 2015). Fundación de Cerro de Pasco. Recuperado de https://www.lhistoria.com/peru/cerro-de-pasco
Hurtado V, Ronny J, Hernández M, José J (2012) Los conflictos ecológico-distributivos en Puntarenas: el caso de la Mina Bellavista de Miramar. Una aproximación inicial, octubre
Inspira Perú (s.f). Personajes de Perú—Pasco. Recuperado de https://inspiracion.ciep.edu.pe/personajes-de-peru/pasco/
Instituto de Geología, Minería y Metalurgia—INGEMMET, Ministerio de Energía y Minas Geología del Cuadrángulo de Cerro de Pasco Hoja 22-k Lima, Perú, 2011
La Oroya (2011) Urbanismo, arquitectura y patrimonio industrial. Tesis de licenciatura en Arquitectura, Universidad Nacional de Ingeniería. Lima. Minera Yanacocha 2009. Reporte de sostenibilidad
León C (2006) Indicadores ambientales y retos conceptuales para operar políticas públicas, en I. Pisanty y M. Caso (comps.) Especies, espacios y riesgos: Monitoreo para la conservación de la biodiversidad, México, INE
MINSA (2015) Oficina General de Planeamiento y Presupuesto. Cuentas Nacionales de Salud: Perú 1995–2012. MINSA, Lima. http://liferiverphy.eu/Noticias/2015_02_24_Murcia%20enclave%20ambiental_Plantas%20devoradoras%20de%20metales%20pesados.pdf
Ministerio de Transportes y Comunicaciones (2005) Provías Departamental, Banco Interamericano de Desarrollo, Gobierno Regional de Pasco. Plan Vial Departamental—Pasco Informe Final. Cerro de Pasco, Perú
Modelo para orientar la gestión hacia la Conservación y el Desarrollo Sostenible de una Región con Reserva Comunal (2016) Documento de trabajo 22. Lima, Perú, 36 pp
Mundial B (2012) Wealth and sustainability: the environmental and social dimensions of the mining sector in Peru. Main report (2), 194 pp
Municipalidad Provincial de Pasco (2006) Ministerio de Vivienda, Construcción y Saneamiento. Actualización del Plan de Acondicionamiento Territorial de la Provincia de Pasco. Lima, Cerro de Pasco, Perú
Nardin Daniel MCCJ (eds) (1993) Siempre hay un mañana. Yanacancha en la historia de Cerro de Pasco, Editorial Sin Frontera, Lima
Ochoa G (2021) Rigoberto y Ochoa de la Torre José Manuel. Servicios de energía y habitabilidad en los hogares de Sonora, México, ante el Covid-19, Viviendas y Comunidades Sustentables, México, No 10
Pajuelo R (2005) Medio ambiente y salud en La Oroya. Cooperacción, Lima, PUIG
Programa de Adecuación y Manejo Ambiental, Unidad de Producción Cerro de Pasco, agosto
Proderm, Gobierno Regional Pasco (2024) Plan de Acción Ambiental Regional Pasco, Lima, Perú
Rodriguez Larrain S (2019). Evolución de las técnicas constructivas en la vivienda de la comunidad alpaquera De Orduña, Puno, Perú. En 19° Seminario Iberoamericano de Arquitectura y Construcción con Tierra, pp 492–501. FUNDASAL/PROTERRA. Recuperado de http://ibomex.org/archivos/memorias/SIACOT2019_Memorias_completo.pdf
Rueda S (2001) Modelos e indicadores para ciudades más sostenibles en Workshop: Indicadores de huella y calidad ambiental urbana – Fundación Forum Ambiental. Barcelona (publicación interna)
Sánchez-Vázquez L, Espinosa-Quezada M, Riofrio ME (2016a)
Sánchez-Vázquez L, Espinosa-Quezada M, Riofrio ME (2016b) Percepción de conflictos socioambientales en zonas mineras: El caso del proyecto mirador en Ecuador. https://doi.org/10.1590/1809-4422ASOC129708V1922016
Sennett R (1975) Vida urbana e identidad personal. Península, Barcelona

Toni (2009) Marca ciudad. Cómo rediseñar para asegurar un futuro espléndido para todos. Paidós, Barcelona

Tromme JP. Etudes des typologies des habitations et espaces publics dans la région de Cerro de Pasco. Memoria de fin de estudios, Université de Liège. Lieja

Volcan Compañía Minera S.A.A. Memoria Anual, 2005 al 2015

Wieser MF, Rodriguez Larrain S, Onnis S (2021) Estrategias bioclimáticas para clima frío tropical de altura. Validación de prototipo en Orduña, Puno, Perú. ESTOA, vol 10, pp 9–19. Recuperado de. https://publicaciones.ucuenca.edu.ec/ojs/index.php/estoa/article/view/3149

Part III
Ecological Environment and Resource Management

Chapter 11
Construction of Green Design Assessment System Based on Life Cycle Theory

Jun Wang, Ke Yang, and Su Xu

Abstract Under the background that carbon neutrality is becoming an important content of global climate action, Green Design, as a method to efficiently promote the realization of this goals, should also be optimized accordingly. The main purpose of this research is to establish assessment indicators covering the whole life cycle of green design, which improve energy efficiency and reduce carbon emissions during design process. This research explains the development and assessment principle of the whole life cycle, then summarizes the current problems and solutions of life cycle assessment. Derived from design principles, we build a systematization green design logic process from four levels, which contain dimensional layer, objective layer, design strategy and design solution. Based on the content mentioned above, introduces the life cycle assessment indicators into the construction of Green Design Assessment. Construct evaluation indicators covering the pre-design stage, product manufacturing stage, using stage and post-processing stage system, corresponding secondary and sub-assessment indicators can be supplemented according to specific products. In summary, a perfect design assessment method will be more conducive to promoting overall energy use efficiency and assisting the early realization of the carbon neutrality.

Keywords Carbon neutrality · Life cycle assessment · Green design · Evaluation system

11.1 Introduction

With global climate change, natural disasters such as heavy rains, droughts, floods and landslides are occurring more frequently, and the warming climate will lead to global sea level rise, ocean acidification and biodiversity loss. To effectively control

J. Wang · K. Yang · S. Xu (✉)
Department of Landscape Planning and Design, East China University of Science and Technology, Shanghai, China
e-mail: xusutju@126.com

C. Yuan et al. (eds.), *Proceedings of 4th International Conference on Resources and Environmental Research—ICRER 2022*, Environmental Science and Engineering,
https://doi.org/10.1007/978-3-031-31808-5_11

global natural disasters, the Intergovernmental Panel on Climate Change (IPCC) has recommended that global warming needs to be limited to a safe threshold of 1.5 °C by the mid-twenty-first century, which means that global greenhouse gas emissions must be controlled and achieving net zero emissions will become a critical global goal (What is carbon neutrality and how can it be achieved by 2050 2021). In September 2020, China made a solemn commitment to the world at the 75th session of the United Nations General Assembly to achieve "Carbon Peaking" by 2030 and "Carbon Neutrality" (Xi Jinping's speech at the general debate of the seventy-fifth United Nations General Assembly 2020) by 2060. In December of the same year, the Central Economic Work Conference proposed to formulate an action plan to achieve "carbon peaking" and support places with the conditions to take the lead in peaking (The central economic work conference held an important speech by Xi Jinping and Li Keqiang 2020). In the work report of the two sessions of the government in 2021, it was further clarified that the goal of "carbon neutrality" should be ensured and promoted through a series of measures such as new energy, innovation and institutional construction (Government work report 2021). In the trend of achieving "carbon neutrality" and comprehensive green transformation of the economy, green design provides a development direction for the realization of this goal. Green design upholds the concept of sustainable development and ecological protection, expands the use of green energy, and clearly systematizes the design system. Green design requires that the whole process of design must be strictly in accordance with the indicators of environmental protection to implement the production of each stage, therefore, the whole process of design production and use will be predicted, and the overall energy consumption will be more accurately predicted and optimized, which is more conducive to promoting energy saving and emission reduction, and improving the overall energy efficiency, and in this context, it is particularly important to build a green design assessment system based on the whole life cycle perspective.

11.2 Methods

11.2.1 Review the Development of Life Cycle Assessment

Life cycle assessment (LCA) was developed in the 1970s as a method to assess all phases of a product from production to recycling, including the acquisition of raw materials, product design, manufacturing, transportation, use, maintenance, and post-recycling (Ishii 1998).

This concept was first applied to the environmental impact assessment of Coca-Cola's packaging. As early as 1969, the Midwest Resource Institute conducted a quantitative analysis of Coca-Cola's packaging in order to identify the different environmental impacts of plastic and glass packaging materials, and the results of this study prompted Coca-Cola to switch to plastic as a more environmentally friendly form of product packaging (Curran 1993). Most of the research in this period was oriented towards the packaging of products, mainly initiated by some companies and factories, and the corresponding research results were used as tools for internal product development and decision making. After 1975, the U.S. National Environmental Protection Agency shifted its research objectives from individual product analysis to energy conservation and solid waste reduction; after 1985, it further shifted from pollutant emissions to energy analysis. In 1985, the Dutch Ministry of Housing, Planning and the Environment (NHPE) proposed for the first time to standardize the basic methods and data for LCA, oriented to all aspects of the product from production and consumption to final waste disposal (Yang and Wang 1998). In 1990, the International Society of Environmental Toxicology and Chemistry (SETAC) first proposed the concept of total life cycle, and then held several academic seminars (Finnveden et al. 2009). In 1993, the International Standard Organization (ISO) started to draft a series of ISO 14000 standards for environmental management and incorporated LCA into them. In 1997, ISO 14040 standard "Principles and Framework for LCA of Environmental Management" was formally promulgated, in which the assessment framework of LCA was further clarified, mainly including four parts: goal and scope definition, inventory analysis, impact assessment, and interpretation of results. LCA is widely used as a standardized scientific management decision-making tool in the fields of production assessment, product optimization, and related policy designation.

Since the 1990s, China has gradually introduced LCA theory-related research, and introduced the ISO 14040 series of standards in 1998, and transformed them into the GB/T 24,040 series of national standards. With the current increase in ecological and environmental protection in China, the government has also issued a number of policy documents to promote the related application of LCA, and the current LCA in China has been widely used in metal smelting, energy measurement, waste treatment and products, construction and other fields (Zhai et al. 2021).

The design assessment of early products was only oriented to part of their life cycle, i.e., from the supply of raw materials to the loss of function and disposal of the product, and the overall process was an open-loop one, considering only the basic attributes of product use, i.e., from "Cradle-to-Gate". In contrast, the whole LCA is an extension of the life cycle of a commonly designed product at both ends, and the systematic strategy fully considers the environmental factors of the whole process from raw material acquisition to final disposal, forming a closed loop of energy reuse flow. The time span and depth of the study are largely expanded, and the scope, data quality, and methodological results of the study are transparent (Su and Wang 2008). Given that all stages of product production have an energy input–output relationship with the physical environment, a systematic and global assessment is necessary for manufactured products in the closed-loop process of preconceived design, market research, technology development, raw material acquisition, processing and manufacturing, transportation and storage, installation and use to later product maintenance, waste recycling and reuse, and thus whole LCA is also referred to as a "Cradle-to-Grave analysis", i.e., the entire life cycle of energy input and output from manufacturing to disposal (Keoleian et al. 1994). The LCA approach extends the energy assessment of the product upfront and the energy reuse of the subsequent process, extending its energy and use efficiency beyond material acquisition and waste management systems to form a special "Cradle-to-Cradle" assessment model, i.e., expanding to generate new energy use from waste recycling (Birgisdottir et al. 2007; Jian-fei and Li-hong 2004). In the case of pre-production products, LCA can help to minimize and prevent foreseeable future environmental problems at the source and maximize energy utilization; in the case of finished products, the indirect environmental impacts caused by end-of-life can sometimes exceed the direct impacts of the waste management system itself, and the broad perspective of LCA can help to obtain greater environmental and even social benefits through different waste management processes again, linking the energy reuse in the production process from product disposal to raw material supply, providing clear and appropriate requirements for the production process through analysis, integrating the user's intention requirements to give optimal technical implementation, and at the same time accommodating new scientific discoveries and technological developments, ultimately achieving sustainable development of manufacturing at the source (Zhang et al. 2021).

11.2.2 The Principle of Whole LCA

Product LCA is identified and quantified through the use of materials and energy and the resulting environmental pollution emissions, and is divided into four main phases, namely, the definition of objectives and scope, Life Cycle Inventory analysis (LCI), Life Cycle Impact Assessment (LCIA), and interpretation of results (Fig. 11.1).

① Goal and Scope Definition (GSD)

The definition of goal and scope refers to the detailed qualitative description and quantification of the services expected to be provided by the finished product by

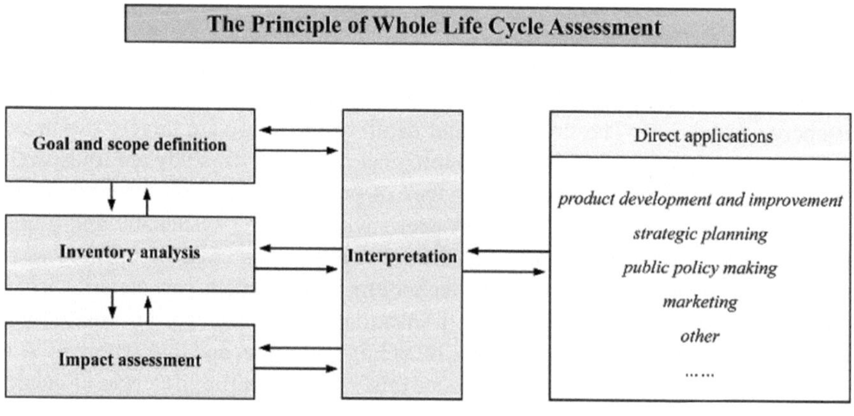

Fig. 11.1 Whole life cycle assessment process

defining the goals and intended uses according to the product system boundaries (Hauschild et al. 2005).

② **Life Cycle Inventory Analysis (LCI)**

Life cycle inventory analysis refers to the process of objectively quantifying the energy and raw material requirements and environmental emissions throughout the life cycle of the target manufactured product, including the input and output of environmental interactions such as exhaust gases, wastewater, environmental releases, and solid waste, which is highly relevant in general. The material flow of the process requires a large amount of data, and therefore a complete database is needed to record the input and output data of each production step (Muralikrishna and Manickam 2017).

③ **Life Cycle Impact Assessment (LCIA)**

Life cycle impact assessment is the core of LCA and the most difficult part of it. LCIA refers to the qualitative or quantitative assessment of the environmental impact pressures identified in the inventory phase, and the determination of the degree and importance of the impact on the external environment caused by the exchange of material and energy in the production process (Hunkeler and Rebitzer 2005).

④ **Interpretation of Results (Interpretation)**

Interpretation of results refers to the systematic assessment of energy consumption use of raw materials and energy exchange released by interaction with the environment throughout the life cycle of the production process. This step is based on the assessment of the previous objectives and scope to further search for the technological route with the least impact on the environment, and at the same time to be able to make reasonable judgments about the inspection and maintenance of the production process in time to give recommendations and conclusions appropriate to the objectives and scope.

The whole LCA can make decision and assessment on production cost and process based on real-time input and output information, overcoming the one-sided disadvantages of traditional assessment, helping to identify the consumption of resources and energy and environmental emissions at each stage of the production process, accurately discovering every possible opportunity for improvement in the production process, clearly stating the degree of energy input and output at different stages, assisting production enterprises to rationalize and maximize resource allocation, achieve dynamic and timely cost control, thus creating more profits for the company, reducing the pollution that may be controlled for the environment, and finally achieving sustainable production.

Relatively speaking, for the whole LCA, the beginning of all assessment originates from the accuracy of the initial data, so the establishment of the cost database becomes the basis and key of the assessment. The current LCA methods still have some shortcomings, such as old methods, high uncertainty of source data with variability, relatively lack of timeliness, and incomplete data and processes (Xiao and Yang 2018). In LCI, the analysis of the inventory pays great attention to the completeness and reliability of the data sources, timeliness, technical representativeness, and the capacity of the sample, which often requires a large number of human and material resources conditions to support. The selection of the system boundary in the pre-assessment period is usually with subjective factors, incomplete and non-uniform characteristics, and the gaps in data collection still exist, but the higher demand for data quality and completeness, moreover, increases the human, financial resources, time consumption. Because the process is cumbersome, so the assessment has a certain time lag, and when the list data collection is over, it should be strictly audited to check whether the data are accurate, and whether there are omissions, etc. (Liu et al. 2010).

11.3 Discussion on Green Design Assessment Framework Based on Whole Life Cycle

11.3.1 The Logical Process of Green Design

Green design is a broad concept, and there are differences in design strategies and design methods depending on the design object. Therefore, the premise of the assessment of green design is to clarify the logical process of green design, which can be abstracted into several levels because different design objects can appear different from each other in terms of form, function and characteristics.

The first level of green design to be achieved is the dimensional layer. Dimension is the most basic aspect to evaluate a product, for example, for a home appliance product, it can be evaluated in terms of resources, energy, environment and quality. The resource dimension focuses on the consumption of materials and water, the consumption of packaging materials and whether it is easy to recycle; the energy dimension focuses on the assessment of the product in the manufacturing and use of energy conservation and energy efficiency; the environment dimension focuses on the emission of pollutants and release of toxic and harmful substances during the production and use of the product; the quality dimension focuses on the durability, health and safety, and comfort of the product and other aspects.

The second dimension to be achieved by green design is the target layer. The target layer is corresponding to different dimensions, such as the resource dimension mentioned above, and its corresponding target is the amount of consumables, water consumption, recycling rate, etc.; another example is the energy dimension, its corresponding target is the power consumption, energy saving rate, etc.; the environmental dimension corresponds to the target of toxic emissions, unit carbon dioxide emissions, etc.; the quality dimension corresponds to the target of durability, comfort, safety, adaptability, etc. It can be said that the target layer decomposes the content to be emphasized in each dimension, and is the indicator that specifically reflects the dimension. At the same time, it must be noted that different dimensions may produce the same target, for example, the target of water consumption in the resource dimension may also be emphasized by the environment dimension; the power consumption mentioned in the energy dimension also belongs to the resource dimension at the same time, so there is a cross-correspondence between dimensions and targets.

With the dimensions and goals, a third level can be achieved—the design strategy layer. Design strategy is the design thinking and approach to achieve a specific goal, and therefore corresponds to the goal one by one. For example, in order to achieve the goal of reducing raw material consumption, then the design strategy should consider the lightness of the product structure as much as possible, then the structural lightness is a design strategy; another example is the goal of reducing carbon dioxide emissions per unit, it is necessary to adopt possibly more than one emission reduction strategy; another example is to achieve the goal of comfort, then the design strategy may also exist more than one.

Once a design strategy is determined, then different green design solutions can be generated. Often, multiple design strategies may be integrated into a single design solution, and unused design solutions may contain the same design strategy.

At this point, the logical process of green design is constructed, and the process is diagrammed as shown in the following figure (Fig. 11.2).

In the assessment study for the whole life cycle of products, Zhou's life cycle-based sustainable design assessment method is divided into five assessment categories, namely, embedded energy, gray energy, induced energy, operational energy and waste energy in construction and industry. Embedded energy refers to the processing method of raw materials; gray energy refers to the energy consumed in the transportation phase after the product is made; induced energy refers to the utility consumption and pollution emissions generated by the integration of construction after transportation; operational energy refers to the energy consumed in the later use of the building or product, including the maintenance energy consumed by wear and tear;

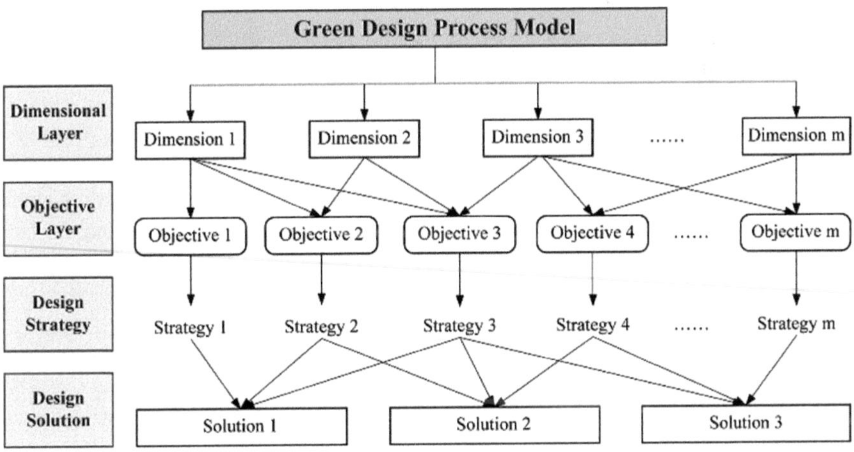

Fig. 11.2 Logical process construct of green design

and disposal energy refers to the energy consumed in the recycling phase after the building or product is disposed of. This classification makes the whole LCA of buildings and industries more complete (Zhou 2020). In the construction of the green food packaging system, Zhao incorporated the dimension of support manufacturing in the manufacturing stage to form a supply chain-based cycle strategy management (Zhao 2020). In the study of electric vehicles, Zhao et al. emphasized the relative nature of evaluating the accounting nightingale of carbon emission reduction by comparing the upfront manufacturing cost of electric products with the later energy consumption in a comprehensive manner, and the consideration of the later use and maintenance of electric vehicles should not be neglected because of their higher upfront energy consumption, i.e., the whole life cycle needs to be analyzed and evaluated in a holistic manner (Zhao et al. 2021). In the reduction design for bamboo furniture, Liu et al. proposed to reduce and reconfigure from the whole life cycle perspective from each link in turn, i.e., the full-domain production planning is then refined into each production stage for optimization (Liu and Yu 2021).

Based on the manufacturing process of the whole life cycle of enterprise products and the assessment suggestions added by the above scholars, the product-oriented whole LCA is divided into four main stages: pre-design, product manufacturing, putting into use and post-processing.

11.3.2 Pre-design Stage

The pre-design stage mainly includes three indicators: design anticipation, market research and technology research and development. Among them, design anticipation includes demand estimation, production planning and product deepening; market research includes competitive product selection, material selection and product process; technology research and development includes structure composition, material acquisition and process development.

In the initial analysis of market demand, the opportunity gap refers to the market circulation of a product due to inadequate design or conditions, or product whose value does not meet consumer expectations or bring more environmental problems, resulting in the product is not accepted by the market, thus forming a market opportunity. The company's grasp of the product market opportunity gap side measures the extent of the value of its product development, selects the appropriate demand forecasts for the production process of the product planning, and further deepens the specific design foreseen problems and details in the design. Market research needs to include the analysis of the current competing products of the same type of preset products, which is conducive to the control of product design and the direction of new technologies, while in the process of researching the market to complete the selection of suitable materials, the selection of materials for the product should be comprehensive consideration of resource issues, in the conditions of renewable resources can be used as far as possible to choose renewable resource materials, environmental protection materials and promote growth, alternative materials, understanding the current production and manufacturing process, and promoting the energy cycle of its reuse. Technology research and development includes the deduction of new product construction, new technology research and development for the acquisition of materials, and research and development for manufacturing processes.

11.3.3 Product Manufacturing Stage

The product manufacturing phase is divided into three parts: raw material acquisition, manufacturing, and support manufacturing. The raw material acquisition includes raw materials, recycled materials and processed materials; the manufacturing includes safety design, simulation and design analysis; the support manufacturing includes technology construction, production technology and collaborative manufacturing.

The raw materials for product manufacturing can be obtained from raw materials, reused from recycled materials, or processed into a material source for product production, which determines the later product structure composition, production process performance judgment and output. Manufacturing encompasses the design, simulation, and design analysis of safety in the production of products. In the process of operation and use of any facilities and equipment, there will be the probability of accidents, so the design should also be optimized to reduce the accident rate within the safe life, the production process of the non-maintainable and repairable parts of the optimal design, the production of human–machine collaboration process, should be reasonable to improve the working environment of the operator, improve the efficiency and quality of work. The calculation of design solutions through simulation technology in manufacturing is a reliable technical guarantee that problems in manufacturing can be detected in time to optimize the solutions. The design analysis includes the estimation of the designed product life, considering the overall service life, durability, damage tolerance and maintenance cost, which is necessary to improve the competitiveness of the company's products, and the estimation of certain service life is also a decisive factor for the high economic efficiency of the whole life cycle. In addition, it should be added that the production process should control the emission of pollutants and sound and light pollution, and account for the environmental impact caused by direct and indirect emissions. Supporting manufacturing includes the need for component construction of products, the corresponding production technology enhancement, and the need for manufacturing of co-products that operate as products.

11.3.4 Putting into Use Stage

The putting into use stage includes three stages of product transportation, sales, and use. Among them, storage and transportation include product packaging, assembly form and transportation means; product sales include product storage, product sales and product tributaries; use stage includes commissioning and assembly, product operation and testing and diagnosis.

Product transportation includes product packaging, product parts assembly form and transportation mode. From product packaging, reduced, and non-toxic packaging design and easy to disassemble packaging methods to improve transport efficiency while also facilitating the recycling of components; good disassembly not only saves time costs for production, transportation and user assembly, but also provides new use value for product reuse; the development of edible material packaging and biodegradable technology is more conducive to reducing the loss of energy in the whole life cycle. Product distribution includes the storage of products at their destination, the sales process, and the retention of expired products. The environmental consumption conditions for the storage of existing products should be included in the assessment system, as well as the cost of electricity used in the sale and the final disposal of expired products. In the use phase, further consideration needs to be given to all aspects of the commissioning and assembly of the product assembly, the overall energy consumption during operation, and the later testing and diagnosis.

11.3.5 Post-processing Stage

The post-processing of products can be divided into product maintenance, recycling and post-disposal. Among them, product maintenance includes performance failure, economic failure and maintenance cost; product recycling includes clean production, recycling and raw material supply; product disposal includes product disassembly, composition analysis and biodegradation.

After the product performance failure as well as economic failure, the product needs to be discarded and maintenance judgment, the product maintenance service costs and other economic aspects of the full range of assessment, to achieve economic optimization. Recycling of products includes clean production, recycling of products and supply of other products as raw materials. The discarded products can be disassembled from the product parts, or its material composition can be precisely monitored and analyzed to achieve the extraction and reuse of microelements, combined with the rapid assessment technology of the current status of waste packaging recycling, green disassembly technology and sorting and recycling technology to fully green the residual waste, and the completely discarded products can be biodegraded.

In addition, when conducting the whole LCA, enterprises also need to keep in mind that product design is not only about maximizing the environmental, economic and social benefits generated by energy, but also to build it on the actual functional needs of consumers, and to enhance energy efficiency while ensuring the market acceptance of the product is the optimal embodiment of the implementation of the whole LCA method.

11.4 Conclusion

The whole LCA method covers the whole process of production and use of vision, effectively achieving the goal of sustainable development of manufacturing industry from the source, and the technology has gradually become a powerful tool for clean production, process improvement and green design and manufacturing of enterprises. Foreign countries have done a lot of research on the whole life cycle evaluation model, while China's research on its late start. In view of the differences in data between different countries and regions, it has become a necessary task to promote the localization and optimization of the whole LCA method, and at present, China's whole LCA in the four stages of goal and scope definition, life cycle inventory analysis, life cycle impact assessment and result interpretation. There are still problems such as incomplete data, inadequate model construction and insufficient software and industrial system, etc. existing in these four stages, so it is still necessary to explore and develop the software and methods of whole LCA that are more in line with national conditions. Green design is a design method that integrates environmental attributes and meets environmental and energy efficiency while ensuring product quality, which plays a crucial role in the sustainable development of manufacturing industry. Integrating the whole life cycle concept into the green design evaluation theory and evaluation system for construction requires us to first clarify the basic logic of green design. Given that different products have different materials, resource consumption, recycling rate, design methods and solutions, it is necessary to establish the corresponding logical system from the four levels of design dimensions, design objectives, design strategies and design solutions of different products to realize the complementarity between each dimension.

The green design assessment framework of the whole life cycle can be divided into four stages: pre-design, product manufacturing, putting into use and post processing, in which design anticipation, market research and technology development belong to the pre-design stage; product manufacturing includes the acquisition of raw materials, production, manufacturing and corresponding support manufacturing; the putting into use stage supplements the indicators during product transportation, sales and use; the post processing covers the maintenance, recycling and disposal of the product. And the corresponding secondary indicators and sub-indicators are established as a benchmark to refine the construction of the supplementary evaluation system, to maximize the consideration of the use of energy efficiency at each step of the process, and to improve the overall efficiency from the source, and it is believed that with the continuous accumulation of practical experience, the construction of the green design evaluation system based on the whole LCA will be mature and more perfect.

Acknowledgements This research is supported by China National Social Science Fund, No. 19BJY063.

References

Birgisdottir H, Bhander G, Hauschild MZ, Christensen TH (2007) Life cycle assessment of disposal of residues from municipal solid waste incineration: Recycling of bottom ash in road construction or landfilling in Denmark evaluated in the ROAD-RES model. Waste Manage 27(8):75–84

Curran MA (1993) Broad-based environmental life cycle assessment. Environ Sci Technol 27(3):430–436

Finnveden G, Hauschild MZ, Ekvall T, Guinée J, Heijungs R, Hellweg S, Koehler A, Pennington D, Suh S (2009) Recent developments in life cycle assessment. J Environ Manage 91(1):1–21

Government work report (2021). http://www.gov.cn/zhuanti/2021lhzfgzbg/index.htm. Last accessed 5 Aug 2021

Hauschild M, Jeswiet J, Alting L (2005) From life cycle assessment to sustainable production: status and perspectives. CIRP Ann 54(2):1–21

Hunkeler D, Rebitzer G (2005) The future of life cycle assessment. Int J Life Cycle Assess 10(5):305–308

Ishii K (1998) Design for environment and recycling: overview of research in the United States. In: CIRP 5th international seminar on life-cycle engineering, pp 16–18

Jian-fei G, Li-hong Q (2004) Product development process modeling with product life-cycle consideration. Comput Integr Manuf Syst 10(1)

Keoleian GA, Glantschnig WJ, McCann W (1994) Life cycle design: AT&T demonstration project. In: IEEE international symposium on electronics & the environment, pp 134–135

Liu XL,Wang HT, Chen J, He Q, Zhang H, Jiang R, Chen XX, Hou P (2010) Methodology and base model of Chinese life cycle reference database. J Environ Sci 30:2136

Liu ZM, Yu GW (2021) Research on the reduction design of bamboo furniture based on the concept of whole life cycle. Furnit Interior Des 3:1–6

Muralikrishna IV, Manickam V (2017) Chapter five—life cycle assessment. Environmental management. Elsevier, Amsterdam, The Netherlands

Su XM, Wang RR (2008) Research on life cycle green assessment of product. J Mod Manuf Eng 07:11

The central economic work conference held an important speech by Xi Jinping and Li Keqiang (2020). http://www.gov.cn/xinwen/2020-12/18/content_5571002.htm. Last accessed 5 Aug 2021

What is carbon neutrality and how can it be achieved by 2050? https://www.europarl.europa.eu/news/en/headlines/society/20190926STO62270/what-is-carbon-neutrality-and-how-can-it-be-achieved-by-2050. Last accessed 5 Aug 2021

Xi Jinping's speech at the general debate of the seventy-fifth United Nations General Assembly. http://www.mofcom.gov.cn/article/i/jyjl/l/202012/20201203020929.shtml. Last accessed 5 Aug 2021

Xiao HX, Yang DH (2018) Methods and applications of environmental impact assessment based on product life cycle: a review. Urban Environ Stud 01:88

Yang JX, Wang RS (1998) Review and prospect of life cycle assessment. Chin J Environ Eng 02:22

Zhai Y, Zhang T, Ma X, Shen X, Ji C, Bai Y, Hong J (2021) Life cycle water footprint analysis of crop production in China. Agric Water Manage 256

Zhang H, Wu SL, Wang ZY (2021) Research on the comprehensive benefits of assembled buildings from a whole life cycle perspective. J Liaoning Univ Technol (Social Science Edition) 23:46

Zhao L (2020) Construction of green food packaging system based on strategic management of supply chain cycle. Food Mach 36:115

Zhao ZX, Shao CF, Chen J (2021) Assessment of the whole-life carbon emission reduction effect of private electric vehicles in China's provinces. Environ Sci Res 1–20

Zhou HM (2020) Sustainable design ideas and methods based on "whole life cycle assessment." Ind Eng Des 2:25

Chapter 12
Research on Multi-objective Optimization of Building Energy Efficiency Based on Genetic Algorithm

Cong Wang and Yanyan Lv

Abstract The development of energy efficiency schemes is not only an important means of controlling a building's energy consumption, it is also a key element of green building evaluation. For developers, the choice and implementation of energy efficiency solutions will also have a direct impact on their economic efficiency. Therefore, it is essential to study how to balance energy efficiency, evaluation scores and economic benefits in the development of schemes. Previous academic research has focused on evaluating the feasibility of established solutions for green buildings, but has neglected to combine the three objectives of energy efficiency, evaluation scores and economic benefits to optimise energy efficiency solutions. This paper constructs an optimisation model for green building energy efficiency projects based on an incremental comprehensive benefit model with its incremental cost and incremental benefit as the objective function and a series of constraints. Genetic algorithms are a more effective way of solving combinatorial optimisation problems than orthogonal experimental methods. Therefore, how to use this method for problem solving is also the focus of this part of the research. For the Pareto solution domain generated by the algorithm, the optimal solution evaluation index 'energy efficiency design', i.e. the energy efficiency per unit incremental cost, is designed to determine the final implementation solution.

Keywords Genetic algorithms · Energy efficient buildings · Multi-objective optimization

C. Wang (✉) · Y. Lv
Shanghai Urban Construction Vocational College, Shanghai, China
e-mail: wangcong@succ.edu.cn

C. Yuan et al. (eds.), *Proceedings of 4th International Conference on Resources and Environmental Research—ICRER 2022*, Environmental Science and Engineering,
https://doi.org/10.1007/978-3-031-31808-5_12

12.1 Introduction

In recent years, various climate disasters caused by the greenhouse effect have occurred frequently, and environmental issues have been put on the agenda of various countries, becoming an important issue that needs to be addressed. China has put forward a sustainable development strategy in this regard, requiring the coordinated development of the environment and the economy. With the introduction of the concept of sustainable development, building energy efficiency has begun to transform into green building. As a new concept in the construction industry, the development of green building in China has been constrained at this stage due to its implementation costs and the benefits that can be obtained. In particular, the uncertainty of incremental costs and incremental benefits makes developers often conservative about developing green buildings. The choice of energy efficiency options will have a direct impact on the incremental costs and benefits, and the ability to effectively assess the economics of green building energy efficiency options will have a bearing on whether developers can make scientific decisions. Therefore, it is important to optimise the choice of energy efficiency options based on minimising incremental costs and maximising incremental benefits.

12.2 Construction of a Multi-objective Optimization Model for Building Energy Efficiency Based on Genetic Algorithm

12.2.1 Genetic Algorithm-Based Multi-objective Optimization Model Construction Process for Building Energy Efficiency

This paper proposes the concept of energy efficiency design efficiency, i.e. maximising the incremental benefit achieved per unit incremental cost, and uses it as an evaluation indicator for the solution domain of the objective function to derive the optimal energy saving solution. The construction process of the energy saving solution optimisation model is shown in Fig. 12.1 (Liu 2022; Yan and Dexuan 2022).

Construction of a Multi-objective Optimization Model for Building Energy Efficiency Based on Genetic Algorithm

Multi-Objective Functions

Determine the objective function. The skill technique used to maximise energy savings is the incremental benefit max Z_1, which is expressed in the Eq. (12.1).

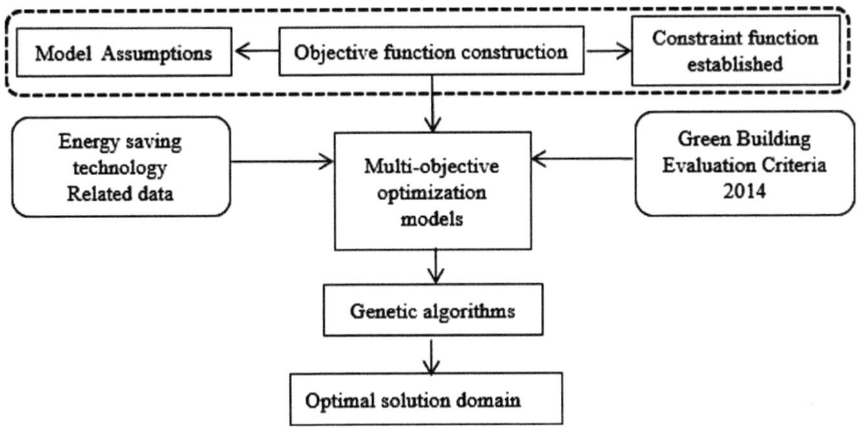

Fig. 12.1 Building energy efficiency solution optimisation model building process

$$\max Z_1 = \Delta S = \sum_{i}^{m} \sum_{j}^{n} \Delta Q_{ij} x_i x_{ij} [P_2 + \delta \cdot \varepsilon + P_3] \cdot P(\Delta S_\alpha, i_0, n) \quad (12.1)$$

The energy saving technology adopted is capable of minimising the incremental cost, and the incremental cost is min Z_2, which is expressed in the Eq. (12.2).

$$\min Z_2 = \Delta C_d + C_c = P_1 + \sum_{i=1}^{m} \sum_{j=1}^{n_i} \Delta C_{ij} x_i x_{ij} \quad (12.2)$$

In the equation:

Z_1 indicates incremental benefit (¥/m^2); Z_2 indicates incremental cost (¥/m^2); ΔS indicates incremental benefit over the calculation period (¥/m^3); ΔQ_{ij} indicates the energy savings $\text{kWh/m}^2 \cdot \text{a}$ generated by the ith energy saving option. P_1 is the price of electricity at the site of the project (¥/kWh); δ is the coefficient of conversion of electrical energy into standard coal, generally we take the value 0.0004. ε is the value of energy savings and emission reductions, taken as 1042.3, which is obtained by adding up the value of pollutant emission reductions, where CO2 is measured at ¥390 per tonne through treatment costs, SO2 is tested at ¥4,344.93 per tonne through the losses caused, and NOX and soot are obtained through energy reports at ¥632 and ¥275 per tonne respectively. P is the discount factor; ΔS_α indicates the annual incremental benefit; i_0 indicates the benchmark rate of return, typically 2–12%, and n is the calculator, both of which values are typically determined by the user. ΔC_d and ΔC_c are the incremental costs during the decision-making period and construction period respectively (¥/m^2); P_2 is the consultation cost during the decision-making period, generally around 0.1% of the construction cost; ΔC_{ij} indicates the incremental cost of the jth energy saving option for the ith technology, in units of (¥/m^2) (Wenjiao 2022; Zhuo et al. 2022; Jiachuan and Mi 2021).

x_i indicates the ith energy saving technology measure.

$$x_i = \begin{cases} 1, & \text{Adopt the ith energy saving technology measure} \\ 0, & \text{No energy-saving technical measures of type i} \end{cases}, \text{ i takes the value of 1, 2, 3, 4, 5}$$

nCorresponding to species technologies such as facade envelopes, air conditioning and heating systems, lighting systems, renewable energy sources and roof energy efficiency, respectively.

x_{ij} indicates the ith energy saving technology measure.

$$x_{ij} = \begin{cases} 1, & \text{Adopt skill technology measure j} \\ 0, & \text{No technical measure of skill j} \end{cases}, \text{ j takes the value of1, 2, 3, 4, 5}$$

Is the jth energy saving option chosen by the decision maker for energy saving technology i. Is the number of energy saving options chosen for energy saving technology i (Yixin and Qiong 2021; Hou et al. 2021).

Energy saving design efficiency. The result obtained after the operation is generally the optimal solution domain. In order for developers to make better decisions, this paper proposes the concept of energy efficiency design, which is the optimal solution evaluation index, i.e. the maximization of the incremental benefit generated by the incremental cost per unit, and can be expressed in Equation max Z, which is expressed as follows.

$$\max Z = \frac{f_{Z_1}(Z_1, Z_2)}{f_{Z_2}(Z_1, Z_2)} \tag{12.3}$$

where: $f_{Z_1}(Z_1, Z_2)$ and $f_{Z_2}(Z_1, Z_2)$ denote the incremental benefit and incremental cost of the incremental benefit of a set of Pareto optimal solutions $f(Z_1, Z_2)$ respectively. By taking the ratio of the incremental benefit and incremental cost of each set of solutions in the Pareto optimal solution domain, the solution that achieves the greatest incremental benefit per unit incremental cost is obtained and is taken as the final energy saving solution (Xiaoxin and Jingyi 2021).

Constraint Functions

Constraints on the number of energy-saving technologies selected for planning and the number of design options. According to the evaluation index of energy-saving projects in the Green Building Evaluation Standard 2014, the selection of energy-saving technologies for green buildings can be controlled. The following formula indicates that the maximum number of energy-saving technologies that can be selected is M, and the number of energy-saving design solutions that can be selected is called, i.e. the maximum number of solutions that can be selected for the ith technology is n (Wenfa and Xinhua 2021).

$$\sum_{i=1}^{m} x_1 \le M \tag{12.4}$$

$$\sum_{j=1}^{n_i} x_{ij} \le N_i \tag{12.5}$$

The link between the selection of energy efficient designs and the design options within them. If for a particular energy saving technology x_i all of the design options within it are not selected, then that energy saving technology must not be selected; if one of the design options is selected, then that energy saving technology must be selected (Huilai et al. 2021).

$$\sum_{j=1}^{n_i} x_{ij} = 0, x_i = 0 \tag{12.6}$$

$$\sum_{j=1}^{n_i} x_{ij} \ge 0, x_i = 0 \tag{12.7}$$

Comprehensive consideration of the way the maintenance structure is constructed. The paper combines the criteria and sets the scoring as follows: if the design solution reduces its building load by 5% compared to the reference building, it will score 5 points; if it reduces by 10%, it will score 10 points (Guojun and Baozhi 2021).

$$a_1 = 5.5\% Q_L \le \sum_{j=1}^{n_i} x_1 x_{1j} Q'_{1j} < 10\% Q_L \tag{12.8}$$

$$a_1 = 10, \sum_{j=1}^{n_i} x_1 x_{1j} Q'_{1j} \ge 10\% Q_L \tag{12.9}$$

In the equation, $\Delta Q'_{1j}$ represents the reduction in building load for design option X_{ij} compared to the reference building, Q_L represents the total load of the reference building and a_1 represents the score obtained for this energy saving option according to the above settings.

Air conditioning and heating systems used in green buildings. The energy consumption of the air-conditioning and heating system is different, and its energy consumption reduction magnitude is also different, and the scoring score will also be different. According to the green building evaluation standard, the scoring method is reasonably simplified in this paper, so as to form the abstraction content, and the mathematical language is refined and expressed, as shown in Table 12.1 (Xie Cunren and Feng 2021).

Where: ΔQ_{2j} is the amount of energy saved by design option X_{2j}, a_2 is the amount of energy used for air conditioning and heating in the reference building and represents

Table 12.1 Air conditioning system energy efficiency evaluation scores

$\sum_{j=1}^{n_i} x_2 x_{2j} \Delta Q_{2j}$ as a percentage of Q_A	a_2	Formula
5–10%	12	(1.10)
10–15%	28	(1.11)
≥15%	37	(1.12)

Table 12.2 Air conditioning system energy efficiency evaluation scores

$\sum_{j=1}^{n_i} x_4 x_{4j} \Delta Q_{4j}$ as a percentage of Q	a_4	Formula
1–1.5%	8	(1.13)
1.5–2%	10	(1.14)
2–2.5%	10	(1.15)
2.5–3%	14	(1.16)
3–3.5%	16	(1.17)
3.5–4%	18	(1.18)
4%≥	20	(1.19)

the score obtained for the energy saving option adopted for the air conditioning and heating system according to the scoring method set out in this paper (Ganghui 2021).

Comprehensive energy use study. This item is mainly based on renewable energy and scored according to the proportion of hot water provided, the proportion of heat and cold and the proportion of electricity. In this paper, by refining and abstracting the evaluation criteria for renewable energy, the scores are allocated according to the amount of energy saved, as shown in Table 12.2.

Where: a_4 indicates the score obtained from the adoption of renewable energy technologies according to the scoring method set out in this paper, ΔQ_{4j} indicates the energy saved by the adoption of renewable energy, and Q is the total energy consumption of the reference building.

Roof energy efficiency evaluation score. Roof construction is also a major factor in the energy consumption of a building, and the energy efficiency of different roof constructions can produce very significant variability. Therefore, the evaluation criteria give different scores depending on their energy efficiency, as shown in Table 12.3.

Green building evaluation control items and scoring items. The control items are generally mandatory standards that traditional buildings must meet, i.e. the Energy Conservation Design Standards for Buildings; the scoring items are the main basis

Table 12.3 Air conditioning system energy efficiency evaluation scores

$\sum_{j=1}^{n_i} x_5 x_{5j} \Delta Q'_{5j}$ as a percentage of Q_L	a_5	Formula
5–10%	5	(1.20)
≥10%	12	(1.21)

for determining the grade of green buildings, requiring that the score for each type of evaluation index must be greater than or equal to 40 points, so the following formula indicates that the score for energy-saving items is not less than 40 points (Rudai et al. 2018).

$$\sum_{i=1}^{m} a_i x_i \geq 40 \tag{12.22}$$

where: a_i has the same meaning as before and indicates the score obtained for energy efficiency measure X_i.

Mutually exclusive energy-saving technology solutions satisfy constraints. There are two types of relationships, parallel and mutually exclusive, between energy saving technologies. When an energy saving technology is designed as a mutually exclusive solution, the following constraints must be met (Miaomiao et al. 2016).

$$\sum_{j=1}^{ni} x_{ij} \leq 1 \tag{12.23}$$

Maximum Incremental Costs. Controlling incremental costs is not only an objective function for investors when making programme choices, it is also a constraint function. Investors usually choose the maximum incremental cost they can afford when investing in a project, which is shown in Eq. 1–24.

$$P_1 + \sum_{i=1}^{m} \sum_{j=1}^{n_i} C_{ij} x_j x_{ij} \leq w \tag{12.24}$$

In the equation, **w** represents the maximum incremental cost acceptable to the investor and the remaining parameters have the same meaning as before.

12.3 Genetic Algorithm Solving Steps and Parameter Setting

12.3.1 Genetic Representation

For the 0–1 variables set out in this paper, the use of permutation coding is a natural and efficient way of coding. Since there are n_j energy saving schemes under each energy saving technology X_i and a total of m energy saving technologies, a chromosome $\sum_{i=1}^{m} n_i$ is set to consist of a gene. Where the jth gene is an integer selected from the mutually exclusive schemes X_{ij}, X_{ij} is the set of schemes contained in the

Table 12.4 Ranking code

x_{11}	x_{12}	...	x_{1n_1}	x_{21}	x_{22}	...	x_{2n_2}	...	x_{m1}	x_{m2}	...	x_{mn_m}

i-th category of energy saving technologies, which are permutation coded as shown in Table 12.4.

12.3.2 Initializing the Population

Completely randomly generated according to the constraint function.

12.3.3 Adaptability Assessment

The selection ranking method is generally used, rather than the objective function value method, with the function represented as follows.

$$Eva(x_i) = \theta(1-\theta)^{i-1}, i = 1, 2 \ldots popsize \tag{12.25}$$

In the equation, θ denotes random elaboration; i denotes chromosome number.

During the operation, the value of θ is first determined, typically $\theta \in (0, 1)$; then the target value is calculated for each chromosome; next, the sorting is done according to the target value; $i = 1$ is set to solve for the fitness value of the ith chromosome and continues until i = popsize (population size).

12.3.4 Determining Roulette as a Selection Operator

The roulette wheel method determines the magnitude of each chromosome's probability of being selected based on the magnitude of its fitness value. The roulette wheel method first calculates the probability of reproduction for each chromosome, i.e. the proportion of each chromosome's objective function value in the sum of the objective function values of all chromosomes in the population; then a random probability of selection is determined.

12.3.5 *Using Arithmetic Hybridisation*

Hybridisation is an important way of population renewal in which the arithmetic process first determines the crossover probability P_c, typically $P_c \in [0.4, 0, 0.99]$; then, a random probability r is generated so that it is between 0 and 1; if the probability r is less than or equal to P_c, x_i is selected as the parent chromosome until each chromosome has performed the operation. Crossover is worked on all the selected parents until a child is produced that can replace the parent to form a new population (Li et al. 2016).

12.3.6 *Determining the Variation Operator*

Variation is an important operation for maintaining population diversity, and similar to crossover, in the calculation process, variation probability P_m is first determined, generally $P_c \in [0.001, 0.5]$; then, a random probability r is generated so that it is between 0 and 1; if probability r is less than or equal to P_m, the direction of variation d and parameter λ are randomly generated so that chromosome x_i is replaced by $x_i + \lambda d$. This step is repeated until each chromosome performs the variation operation.

12.4 Conclusion

The incremental costs and benefits of green buildings have long been a hot topic of academic interest, and it is important for society to have a clear and thorough understanding of them. For homeowners, investment efficiency is the key factor that drives their decisions; for governments, quantified and uniform incremental costs are the cornerstone of their incentives; for consumers, 'value for money' is the scientific criterion for their choices. Energy efficiency projects are the main source of incremental costs and benefits, so it is important to optimise energy efficiency solutions to maximise the incremental benefits per unit of incremental cost. With the development of technology and the increasing variety of building energy efficiency materials and technologies, this paper only takes a limited number of schemes as examples, and there are still shortcomings in the experimental data. In order to simplify the calculation, it is assumed that each scheme is an independent variable and the combination is optimised, thus ignoring the interconnection between the schemes. With the development of computer technology such as big data and cloud computing, it will be possible in the future to make a normative measurement of the energy savings of various materials and technologies and to establish their interlinked energy efficiency impact coefficients.

References

Cunren X, Feng X, Minhao R (2021) Research on multi-objective optimization of construction schedule based on BIM and genetic algorithm. J Eng Manage 35(03):117–122
Ganghui F (2021) Application of BIM genetic algorithm in the optimization of building construction objectives. Ind Archit 51(01):224
Guojun L, Baozhi Y (2021) Genetic algorithm-based simulation of building component location distribution modelling. Comput Simul 38(07):266–270
Hou J, Cunling Q, Zhihong JL, Chenglin M, Guangyi G (2021) Genetic algorithm-based software development for optimizing seismic isolation bearings in buildings. J Chengdu Univ (Nat Sci Edn) 40(03):268–273
Huilai W, Hongwei T, Feng D (2021) Research on the parametric design method of renewable energy utilization potential of low-density housing based on genetic algorithm. Resid Sci Technol 41(07):62–69
Jiachuan W, Mi G (2021) Multi-objective optimization design of office buildings in Gansu towns based on genetic algorithm. Urban Archit 18(35):104–106
Li T, Yolin L, Yang W (2016) Optimal design of green buildings based on genetic algorithm. Build Energy Effic 44(06):53–57+64
Liu B (2022) Multi-objective optimization of the envelope structure of ultra-low energy grassland houses based on NSGA-II and BP neural network. Chongqing Archit 21(10):26–28
Miaomiao H, Yongmei X, Changxing H (2016) Genetic algorithm optimization of operating parameters of an air conditioning system. Fluid Mech 44(12):65–70
Rudai S, Guanan L, Xiaodong X, Tiexiao S (2018) Research on optimal design of envelope structure based on genetic algorithm. Constr Technol 49(02):145–148
Wenfa H, Xinhua H (2021) Multi-objective optimization and multi-attribute decision model in the pre-design stage of green building. Oper Res Manage 30(07):44–49
Wenjiao W (2022) Research on the optimization of green building design based on whole life cycle cost and carbon emission. Home Ind 02:130–132
Xiaoxin W, Jingyi L (2021) Optimization of light harvesting in Hankou riverside residential area based on generative adversarial network (GAN) and NSGA-2 genetic algorithm. Archit Tech 27(09):84–88
Yan Q, Dexuan S (2022) Multi-objective optimization of office buildings in hot summer and cold winter regions based on parametric design. Constr Sci Technol 17:15–18
Yixin T, Qiong H (2021) Multi-objective design optimization of building performance and its application—an example of genetic algorithm. New Archit 05:84–89
Zhuo Y, Mingyuan G, Zhigang Z, Xuejiao C (2022) Research on optimization of centralized electric heating system based on genetic algorithm. Build Therm Vent Air Cond 41(01):1–5

Chapter 13
Artificial Neural Network Model: A Sensitivity Analysis of Coir-Reinforced Soil Cement with Levenberg–Marquardt Back Propagation Algorithm

Marwin B. Pagsuyoin and Dante L. Silva

Abstract The use of environmental wastes in the production of construction materials such as soil cement has been receiving attention for its sustainable development and for the addition of greener materials in the construction industry. A lot of studies show that the mechanical properties of coir can be explored as an effective reinforcement to such materials. The primary purpose of this study is to examine the mechanical properties and relative importance (RI) of coir-reinforced soil cement to its compressive strength. An experimental program was conducted to test the compressive strengths of the soil cement samples. An artificial neural network (ANN) model was established employing the result of the experimental program that will forecast the compressive strength (CS) of soil cement and as assessed through a sensitivity analysis (SA). The analysis was done by two methods namely, Garson's algorithm (GA), and sensitivity index method. Finally, the developed ANN model was validated by comparing its performance with other prediction modelling tools and selected as the best prediction tool to the compressive strength of a coir-reinforced soil cement.

Keywords Coir · Compressive strength · Artificial neural network · Sensitivity analysis

13.1 Introduction

The construction industry has been expanding in most countries due to the demand for more buildings and infrastructures. Consequently, the demand for construction materials is also becoming higher to accommodate the needs in construction (Bahar et al. 2004). Soil cement, as one of the most conventional materials used in construction since the early times, can be produced by combining soil, a small percentage of

M. B. Pagsuyoin (✉) · D. L. Silva
Mapua University, Manila, Philippines
e-mail: mbpagsuyoin@mymail.mapua.edu.ph

C. Yuan et al. (eds.), *Proceedings of 4th International Conference on Resources and Environmental Research—ICRER 2022*, Environmental Science and Engineering,
https://doi.org/10.1007/978-3-031-31808-5_13

cement, and water (Fan et al. 2018). Soil–cement blocks have been widely employed to build load-bearing masonry (Venkatarama Reddy and Gupta 2005). However, the massive use of these conventional materials creates disadvantages not only to the decreasing supply of cement but also to the environment. This concern brought a need for consideration of other materials that can be used to substitute or replace them (Thyavihalli Girijappa et al. 2019).

The utilization of environmental wastes in the production of soil cements has been receiving attention due to its high contribution to the sustainable development of both construction materials and the environment. One of the most used products in agriculture is the coconut fruit. By-products, particularly coir, ends as wastes and recycling them can be a help in contributing to the sustainability of materials (Pham 2016). Its mechanical properties are similar to those of conventional materials used in the production of construction materials that is why it can be explored and experimented in the soil cement mix (Danso 2017).

This research study investigated the mechanical properties and relative importance of coir to compressive strength of soil cement.

The mechanical properties of coir were identified and analyzed. Then, soil cement mixes incorporating it as a reinforcement were designed. From the mix design, samples were constructed, and their compressive strengths were tested. An ANN model was established using the result of the experimental program that will predict the CS of the soil cement and will be evaluated through a sensitivity analysis. The analysis was done by three methods namely, connection weights algorithm, parametric analysis and sensitivity index method. Finally, the developed ANN model was validated by comparing its performance with other prediction modelling tools and selected as the best prediction tool to the compressive strength of coir-reinforced soil cement.

13.2 Method

13.2.1 Materials

The materials that were used in the soil cement mix design are the conventional materials which include the soil, cement, and water. These were added by coir which will serve as an additive to the soil cement mix.

Soil. There will be three types of soil that were used in the soil cement design which were collected from different locations. The particles that passes thru the 4.75 mm sieve were used for the design mixture (ASTM 2007). The collected soils were tested to determine their properties as tabulated in Table 13.1.

Cement. The cement that was used in the mix design is an Ordinary Portland Cement sold in the local market.

Coir. The coir that was used in the mix design came from the local market in the Philippines. This material was washed and air-dried to remove unwanted dirt

Table 13.1 Properties of utilized soils

Property	Parameters	Values		
		Soil A	Soil B	Soil C
Atterberg's limits	Liquid Limit, LL	28	28	28
	Plastic Limit, PL	22.81	24.4	23.77
	Plasticity Index, PI	5.18	3.6	4.23
Proctor test	Maximum Dry Density, MDD	1.424 gm/cc	1.424 gm/cc	1.424 gm/cc
	Optimum Moisture Content, OMC	21.70%	17.40%	19.40%

Table 13.2 Soil cement mixes for sample preparation

Label	Coir content (by weight of soil) (%)
A	0
B	0.25
C	0.50
D	0.75
E	1.00

(Praveen and Kurre 2021). Then, the coir was cut into an average length 50 mm (Danso and Manu 2020).

Water. The water that was used in the mix design is potable water.

13.2.2 Experimental Program

The mix design was based on the proportion of soil cement created from related studies. Cement content is ten percent of soil weight. On the other hand, water content is based on the result of soil compaction test. The calculation coir content is based on their weight of the soil as tabulated in Table 13.2.

13.2.3 Compressive Strength Test

The CS of a soil cement determines its ability to resist loads that may cause it to break and deform. It is a good way to know whether a certain mix design comprising of conventional materials and adding other materials that have similar mechanical properties to undergo first in these tests before deciding to follow the soil cement mix design.

Table 13.3 Type of test and samples needed

Type of test	Sample properties	Sample quantity
Compressive strength test	Cylinder: 150 mm dia × 300 mm height	11 × 3 = 33 samples

In this research study, adding coir to the conventional materials (soil and cement) was studied by testing the compressive strength of the samples created from the different designed soil cement mixes. The 33 prepared samples have undergone curing for 28 days (Raj et al. 2017). After curing, all samples were tested for compressive strength following the standard test method from ASTM D 1633-00.

For the CS test, cylinder samples were constructed with a 150 mm diameter and 300 mm height (ASTM 2007). The quantity and dimensions of the required samples for the test is tabulated in Table 13.3.

13.2.4 Ann

The researcher utilized the software MATLAB to develop the ANN model. In this software, an application called Neural Network Fitting can be used to develop the said model. This application utilizes a multi-layer feed-forward network with sigmoid hidden neurons (HN) and linear output neurons (Kooshkaki and Eskandari-Naddaf 2019). The network was trained with the Levenberg–Marquardt (LM) backpropagation algorithm (BPA) (Laroza Silva, DANTE, Marcelo De Jesus 2020).

The neural fitting tool used the weight percentage of the cement, coir, and water as the independent variables or input data and the result of the compressive strength test as the dependent variables or the target data.

For the preparation of the network data, each input and target datum were randomly arranged and normalized from −1 to 1 using the equation

$$y = [2(x - x_{min})/(x_{max-})] - 1 \quad (13.1)$$

where x is the datum of a parameter, x_{max} is the maximum value in that parameter and x_{min} is the minimum value in that parameter. The randomized and normalized data was then used in the neural fitting tool.

The training samples were submitted to the model during training, and the model were changed based on its inaccuracy. The validation samples were employed to evaluate the generalization of the model and to terminate training when generalization enhancing were halted. In addition, the testing samples had no influence on the training, providing an independent measurement of model capability throughout and after the training.

Utilized four neurons for input layers (IL), a hidden layer with an undetermined number of neurons, and one neuron for the output layer (OL) while designing the network architecture. The model utilized a tangent-sigmoid function from the IL to the hidden layer (HL), followed by a hyperbolic tangent-sigmoid function from the HL to the OL. Levenberg–Marquardt training algorithm was utilized as the parameter for the training function (Sharifi and Hosainpoor 2020).

Various number of HN were analyzed and tabulated to determine the best neural network. The most suitable model is the one with the highest Pearson's Correlation Coefficient (R) and the lowest Mean Squared Error (MSE) (Silva et al. 2021). The best neural network was identified after several trials to be 4-H-1, where H is the number of neurons in the HL. Hence, the final ANN Model.

13.2.5 Sensitivity Analysis

This phase of the research study utilized the ANN model developed in the previous phase (Macariola and Silva 2019). The relative importance and behavior of soil, cement, coir, and water to the CS of the soil cement sample were evaluated through a sensitivity analysis. The two methods that were utilized in this research study are Garson's algorithm (Cabuñas and Silva 2019) and sensitivity index method.

13.3 Results and Discussion

13.3.1 Descriptive Statistics of the Parameters

The descriptive statistics of the parameters utilized in the study are shown in Tables 13.4 and 13.5. The skewness of the datasets for the input parameters including soil content, cement content, coir content, and water content were observed to be negative which implies that the data array are negatively skewed or skewed left and that the left tail is longer (Cain et al. 2017). For the output parameter, the compressive strength has a positive skewness, which suggests that the data array is skewed right and that the right tail of the distribution is longer than the left. The kurtosis of the datasets for all the parameters was less than −1, which implies that the distribution of the datasets was too flat (Kong and Wang 2020).

Table 13.4 Descriptive statistics of the parameters used in the study Part 1

	N	Range	Maximum	Minimum
Soil content (kg)	40	3.00	27.00	24.00
Cement content (kg)	40	0.30	2.70	2.40
Coir content (kg)	40	0.24	0.27	0.00
Water content (kg)	40	0.54	5.24	4.70
Compressive strength (MPa)	40	0.48	1.52	1.04

Table 13.5 Descriptive statistics of the parameters used in the study Part 2

	Mean	SD	Skewness	Kurtosis
Soil content (kg)	25.625	1.390	−0.182	−1.888
Cement content (kg)	2.563	0.139	−0.182	−1.888
Coir content (kg)	0.131	0.088	−0.293	−1.239
Water content (kg)	5.063	0.223	−0.865	−1.095
Compressive strength (MPa)	1.249	0.160	0.329	−1.053

13.3.2 Develop an ANN Model for Predicting the CS of a Soil–Cement Samples

The core attributes of the developed model consist of the LM as the training algorithm (Du and Stephanus 2018), and the hyperbolic tangent sigmoid as the transfer function (Drisya et al. 2021). The performance of the models upon having a different number of hidden neurons is shown in Table 13.6, while the structure of the model is presented in Fig. 13.1.

Table 13.6 Artificial neural network model simulation result

Hidden Neurons	R Training	R Validation	R Training	R All	MSE
1	0.94606	0.91372	0.91068	0.93634	0.00462
2	0.94414	0.94605	0.96203	0.94186	0.00367
3	0.96740	0.98360	0.94956	0.96503	0.00151
4	0.95816	0.99470	0.99032	0.96666	0.00082
5	0.96332	0.98987	0.98586	0.96708	0.00076
6	0.96682	0.99569	0.97999	0.97112	0.00052
7	0.96662	0.99232	0.99966	0.97203	0.00047
8	0.96416	0.99727	0.95008	0.97237	0.00025
9	0.96779	0.99476	0.99914	0.97399	0.00021
10	0.97001	0.96653	0.99881	0.97268	0.00024

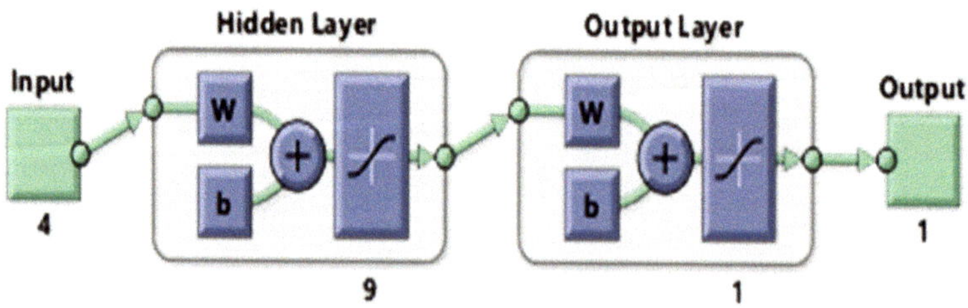

Fig. 13.1 Neural network model structure

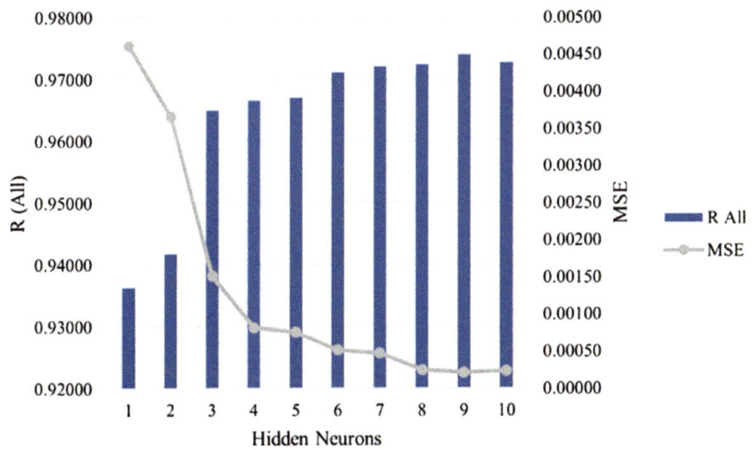

Fig. 13.2 Network performance of varying hidden neurons

The results show that the best model for forecasting the CS of a soil–cement sample is nine (9) hidden neurons, which is consistent with the empirical formula 2 m + 1, wherein "m" is the number of HN (Yan et al. 2020). The performance of the networks upon having varying values of hidden neurons were shown in Fig. 13.2.

The best network has the highest R_{all} and the least MSE. The regression plots of the different phases of development of the network including training, validation, and testing were shown in Fig. 13.3.

13.3.3 Determine the RI of the Input Parameters Using GA

Using the connection weights of the best model, the RI of the input parameters to the CS of a soil–cement sample was obtained utilizing GA (Zhang et al. 2018). The product of the connection weights was presented in Table 13.7.

A bar graph showing the relative importance of the soil content, cement content, coir content, and water content is presented in Fig. 13.4. It was seen that cement is the most dominant parameter in the CS of a soil–cement sample.

The SA using the sensitivity index method was also performed to check the results of the Garson's algorithm. (Silva et al. 2020) The Sensitivity index approach was implemented by eliminating input variables that had no substantial impact on the performance of the model. The sensitivity index values of input variables were calculated by dividing the network's error in the absence of a variable by the network's error with all input parameters. The higher the sensitivity index value corresponds to the more variable significance (Shayannejad et al. 2017). The sensitivity index values of each parameter are shown in Table 13.8. The sensitivity index values obtained confirm the values obtained using the Garson's Algorithm.

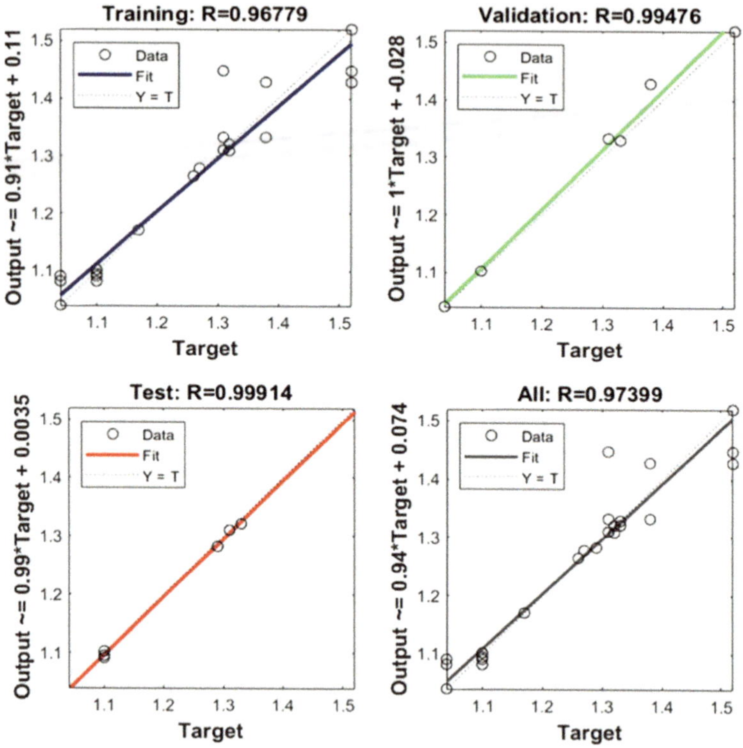

Fig. 13.3 Regression plots of the different phases of neural network development

Table 13.7 Product weights used in the sensitivity analysis

	Soil	Cement	Coir	Water
Hidden 1	0.00908	0.29332	0.36822	0.32938
Hidden 2	0.36363	0.28237	0.29133	0.06267
Hidden 3	0.31215	0.31055	0.06515	0.31215
Hidden 4	0.36817	0.50428	0.05143	0.07612
Hidden 5	0.26695	0.29189	0.16999	0.27117
Hidden 6	0.17378	0.27460	0.36035	0.19128
Hidden 7	0.39419	0.32613	0.16385	0.11584
Hidden 8	0.31538	0.37265	0.15555	0.15643
Hidden 9	0.05769	0.32937	0.28987	0.32306
Sum	2.26101	2.98515	1.91574	1.83810
Relative Importance %	*25.12238*	*33.16834*	*21.28597*	*20.42331*

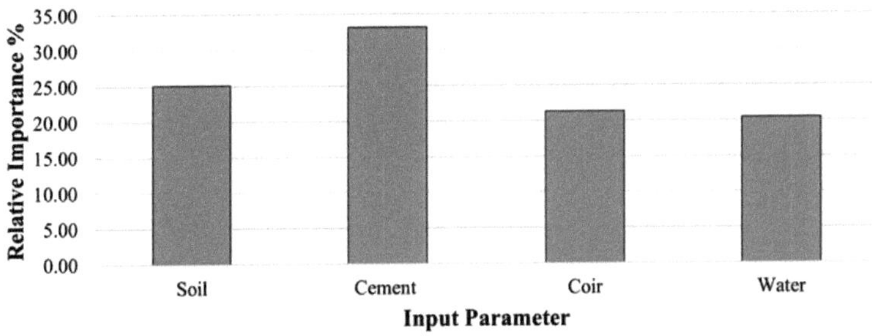

Fig. 13.4 Relative importance of input parameters to the CS of the soil–cement sample

Table 13.8 Sensitivity index analysis results

Removed parameter	MSE	Sensitivity index
No removed parameter	0.00021	1.00000
Soil	0.00033	1.53078
Cement	0.00040	1.87767
Coir	0.00029	1.37765
Water	0.00016	0.74152

13.3.4 Comparison of the Developed ANN Model to Multiple Linear Regression

The developed artificial neural network model was compared with that of a prediction model developed using the multiple linear regression technique. The best model developed using ANN has a mean absolute percentage error (MAPE) of 1.47% while the model created using the multiple linear regression approach has a mean absolute percentage error of 4.71%. The prediction capability of the ANN and MLR models is exhibited in Fig. 13.5. The findings suggest that the developed ANN model is superior to the created MLR models, which is similar to the results of the study by Rao et al. (2021).

13.3.5 Comparison of Developed ANN Model and Other Prediction Modelling Techniques

The developed artificial neural network model is likewise compared with other prediction modelling techniques such as Ensemble of Trees, Regression Trees, Linear Regression, Support Vector Machine, and Gaussian Process Regression. It was seen from the results that the ANN model is the best model as compared to the other prediction modelling approaches created. The performance of each prediction modelling approach checked is displayed in Fig. 13.6.

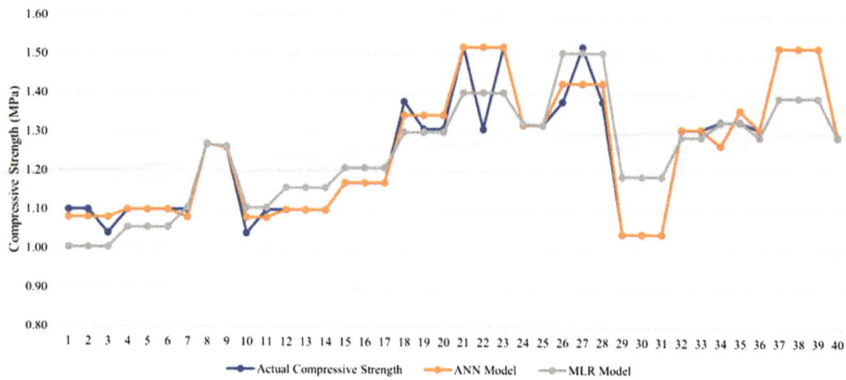

Fig. 13.5 Comparison of model performance–ANN versus MLR

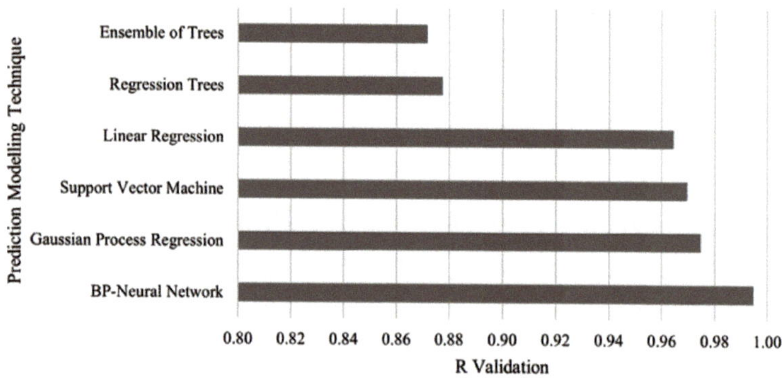

Fig. 13.6 Comparison of model performance with other prediction modelling techniques

13.4 Conclusion

The addition of coir to the soil cement caused a significant increase to the strength. The results showed that the addition of 1.0% coir by weight of soil can produce a coir-reinforced soil cement with the highest increase to its compressive strength. The result of the ANN model simulation shows that the best performance can be generated using nine (9) hidden neurons which resulted to the highest $R_{all} = 0.97399$ and the lowest MSE = 0.00021. The RI of the input parameters to the CS of a soil–cement sample confirms that cement is the most important parameter, followed by soil, coir, and water. The best ANN model that was developed is superior when compared with the created MLR model, as well as with other prediction modelling techniques.

Acknowledgements The authors would like to acknowledge Mapua University, and Department of Science and Technology—Engineering Research and Development for Technology for their valuable supports in this study.

References

ASTM (2007) Standard test methods for compressive strength of molded soil–cement cylinders

Bahar R, Benazzoug M, Kenai S (2004) Performance of compacted cement–stabilised soil. Cement Concr Compos 26(7):811–820

Cabuñas JT, Silva DL (2019) Exploratory factor-item analytic approach for construction project cost overrun using oblique promax rotation for predictors determination. Int J Innov Technol Explor Eng 8(6):47–54

Cain MK, Zhang Z, Yuan KH (2017) Univariate and multivariate skewness and kurtosis for measuring non-normality: Prevalence, influence and estimation. Behav Res Methods 49:1716–1735

Danso H (2017) Properties of coconut, oil palm and bagasse fibres: As potential building materials. Procedia Eng 200:1–9

Danso H, Manu D (2020) Influence of coconut fibres and lime on the properties of soil–cement mortar. Case Stud Constr Mater 12:e00316

Drisya J, Kumar DS, Roshni T (2021) Hydrological drought assessment through streamflow forecasting using wavelet enabled artificial neural networks. Environ Dev Sustain 23:3653–3672

Du YC, Stephanus A (2018) Levenberg–Marquardt neural network algorithm for degree of arteriovenous fistula stenosis classification using a dual optical photoplethysmography sensor. Sensors 18(7):2322

Fan J, Wang D, Qian D (2018) Soil–cement mixture properties and design considerations for reinforced excavation. J Rock Mech Geotech Eng 10(4):791–797

Kong SC, Wang YQ (2020) Formation of computational identity through computational thinking perspectives development in programming learning: A mediation analysis among primary school students. Comput Hum Behav 106:106230

Kooshkaki A, Eskandari-Naddaf H (2019) Effect of porosity on predicting compressive and flexural strength of cement mortar containing micro and nano-silica by multi-objective ANN modeling. Constr Build Mater 212:176–191

Laroza Silva, DANTE, Marcelo De Jesus KL (2020) Backpropagation neural network with feature sensitivity analysis: pothole prediction model for flexible pavements using traffic and climate associated factors. In: 2020 the 3rd international conference on computing and big data, pp 60–67

Macariola RN, Silva DL (2019) Coping with the information age: Development of A data flow diagram-based knowledge management system for mitigating delays for construction. In: IOP conference series: materials science and engineering, vol. 652(1). p 012070, IOP Publishing

Pham LJ (2016) Coconut (Cocos nucifera). In: Industrial oil crops, pp 231–242, AOCS Press

Praveen GV, Kurre P (2021) Influence of coir fiber reinforcement on shear strength parameters of cement modified marginal soil mixed with fly ash. Mater Today: Proc 39:504–507

Raj S, Mohammad S, Das R, Saha S (2017) Coconut fibre-reinforced cement-stabilized rammed earth blocks. World J Eng 14(3):208–216

Rao VV, Garg T, Datta SP (2021) Predictive assessment from ANN and MLR models to optimize the ideal evaporative/hybrid cooler based on experimental observations. J Build Eng 44:103256

Sharifi Y, Hosainpoor M (2020) Compressive strength assessment of concrete containing metakaolin using ANN. J Rehabil Civ Eng 8(4):15–27

Shayannejad M, Eslamian S, Singh VP, Ostad-Ali-Askari K, Shayannejad M, Kazemi M (2017) Evaluation of groundwater quality for industrial using GIS in mountainous region of Isfahan province, Koh-payeh, Isfahan Iran. Int J Constr Res Civ Eng (IJCRCE) 3(3):24–37

Silva D, Villaverde B, De Jesus KL, Marcial ER Jr, Villa-Real CV, Zarrage JM (2020) Design initiative implementation framework: A model integrating Kolmogorov–Smirnov in sustainable practices for triple-bottom-line principles in construction industry. Civ Eng Archit 8(4):599–617

Silva DL, de Jesus KLM, Adina EM, Mangrobang DV, Escalante MD, Susi AM (2021) "Prediction of tensile strength and erosional effectiveness of natural geotextiles using artificial neural network."

In: 2021 13th International conference on computer and automation engineering (ICCAE), pp 121–127

Thyavihalli Girijappa YG, Mavinkere Rangappa S, Parameswaranpillai J, Siengchin S (2019) Natural fibers as sustainable and renewable resource for development of eco-friendly composites: A comprehensive review. Front Mater 6:226

Venkatarama Reddy BV, Gupta A (2005) Characteristics of cement–soil mortars. Mater Struct 38:639–650

Yan C, Li M, Liu W, Qi M (2020) Improved adaptive genetic algorithm for the vehicle insurance fraud identification model based on a BP neural network. Theoret Comput Sci 817:12–23

Zhang Z, Beck MW, Winkler DA, Huang B, Sibanda W, Goyal H (2018) Opening the black box of neural networks: methods for interpreting neural network models in clinical applications. Ann Transl Med 6(11)

Chapter 14
From Food Security to Thinking About the Conservation of Arable Land in China

Xingmei Huang, Gengjie Zhang, Honggang Zheng, Huijie Yang, and Shixin Liu

Abstract With uncertainty over food security due to epidemics, climate change, deteriorating natural resources, and domestic soybean production, arable land conservation was used as an entry point to illustrate how could address this challenge. This paper used literature analysis, data search, and induction to elaborate on the problems in food security and protecting arable land and propose measures to address them. The current issues of arable land conservation are divided into the shortage of arable land per capita, degradation of arable soils, and the situation of 'non-foodisation,' 'non-agricultural,' and 'abandonment of farmland.' To cope with them, measures such as engineering and proposing policy perspectives guarantee food security in China.

Keywords Food security · Conservation of Arable land · Problems · Measures

---

X. Huang · H. Yang · S. Liu
College of Resources and Environment, Yunnan Agricultural University, Kunming 650201, Yunnan, China

G. Zhang (✉) · H. Zheng
Land Resources Science and Technology Engineering Center, Yunnan Agricultural University, Kunming 650201, Yunnan, China
e-mail: jiegeng@ynau.edu.cn

Scientific and Technological Innovation Team for Optimizing the Land Allocation and Ecological Remediation in Southwest China's Diverse Regions, Ministry of Natural Resources, Kunming, China

Field Scientific Observation and Research Bases of Land Use in Yunnan Luliang, Ministry of Natural Resources, Kunming, China

Engineering Laboratory of Land Resources Utilization and Protection in Yunnan Province, Kunming, China

C. Yuan et al. (eds.), *Proceedings of 4th International Conference on Resources and Environmental Research—ICRER 2022*, Environmental Science and Engineering,
https://doi.org/10.1007/978-3-031-31808-5_14

14.1 Introduction

According to the *World Population Prospects* projections report published by the United Nations Department of Economic and Social Affairs in 2019, 'The global population is expected to increase from 7.7 billion in 2019 to 8.5 billion in 2030 and 9.7 billion in 2050, respectively. Expected the global population to grow slowly to 10.9 billion by the end of the century.' (World Population Prospects projections report 2019) Based on this data analysis, the global demand for food is also expected to increase further. The need for food is closely linked to the issue of food security. The Food and Agriculture Organization of the United Nations (FAO) defined 'food security' at the World Food Conference in Rome in November 1974 as 'food security is fundamentally a fundamental right to life for humanity today.' Not meeting this basic right to life can easily lead to socio-political unrest, child malnutrition, and other problems. Arable land is the basis for national food security (Fan et al. 2018), and it is essential to protect it to ensure a steady increase in food production and to meet the food needs of the people. This paper will provide a comprehensive discussion of the conservation of arable land.

14.2 Issues Facing Food Security in Both China and Overseas

The impact of COVID-19, climate change, and the deterioration of natural resources have severely challenged food security in both China and overseas. Including the emergence of declining world food stocks and supplies, higher food prices, increased CO_2 concentrations affecting crop quality, and the deterioration of natural resources affecting crop yields.

14.2.1 The State of International Food Security

Impact of COVID-19 on Food Production and Trade. The global outbroke of COVID-19 has led to a decline in food availability and storage in many countries, a contraction in food trade and an increase in world food prices, and an imbalance in the food supply (Fu 2021; Huang et al. 2021). Reduced food exports from Southeast Asia countries, such as Singapore, banned the export of over 7 million tonnes of food crops (Wang et al. 2021), Rose food prices, and supplied shortages in some countries and regions (Wang et al. 2020a). A report on the website of the Japanese magazine *Nikkei Asian Review* quoted the Thai Business Council as saying that the benchmark export price of Thai rice had risen to $550 (RMB 3,896) per tonne. Argentina's government issued Decree 230/2020 to increase the export tax rate on soybeans and soy products to 33% (Cui et al. 2020). India, one of the world's top five rice exporters, rice exports fell from 14.4 million tonnes in 2019 to 13 million tonnes in 2020 due to COVID-19 (Fu 2021).

Food Issues Caused by Climate Change. Changes in crop growing season due to warming, heavy precipitation, weather extremes; unstable regional water supplies, reduced crop yields, and increased atmospheric CO_2 concentrations can reduce crop nutrient content (Xu et al. 2020; Chen et al. 2021). A paper published in *Nature Climate Change* by Harvard researchers shows that increasing atmospheric CO_2 concentrations from 400 to 550 ppm will reduce crops' protein, iron, and zinc levels by 3–17%. As temperatures rise, prolonged periods of no or low precipitation lead to drought, which can efficiently reduce crop yields. In 2015–2016, the impact of El Niño triggered drought in Ethiopia due to reduced rain, resulting in crop failure (Anderson et al. 2019). Wheat production in India decreased by 5.2% from 1981 to 2009 due to higher temperatures (Chen et al. 2021). Soybean prices rose to over $14 per bushel on CBOT in 2021 due to high temperatures and drought in Argentina due to La Niña (*China Agriculture Outlook Report (2021–2030)*). Still, excessive rainfall in Brazil resulted in lower soybean production.

The Deterioration of Natural Resources Cannot Support Human Needs for Food Security. On 9 December 2021, the FAO released the *State of Land and Water* 2021 report (*State of Land and Water* 2021 report), which stated that 'future food security will depend to a large extent on the conservation of land, soil, and water resources.' But the state of the planet's land, soil, and water resources continued to deteriorate: anthropogenic soil degradation affected 34% of agricultural land (1.66 billion hm^2), there was little quality land available for development, and rapid urban expansion was affecting land and water resources, meanwhile jeopardising the very foundations of food security, and the area of land per capita has decreased by 20% in the 17 years from 2000–2017. The deterioration of the condition of the land, soil, and water, which are critical natural resources for food production, will have a severe impact on food production, including yield and quality, and ultimately on global food security.

14.2.2 The State of Food Security in China

Chinese High External Dependence on Soybeans. Of Chinese major food crops (soybeans, rice, wheat, and corn), soybeans had the highest external reliance (Su and Ji 2020). According to the *Chinese Agricultural Outlook Report (2021–2030)* (*China Agriculture Outlook Report (2021–2030)*), Chinese soybean imports exceeded 100 million tonnes in 2020 and are expected to be 102.5 million tonnes in 2021. In this paper, we selected the relevant data of soybean, corn, wheat, and rice (including rice and early rice) in China in 2019 (Table 14.1) and the total population of 141.08 million at the end of the year for calculation. By calculating the area of arable land required to grow each major food crop and the area of arable land per capita, we clarified the reasons for China's high external dependence on soybeans in terms of arable land. If China produces all the above major food crops, the calculation shows

Table 14.1 Data on major food crops in China in 2019

Major food crops	Production (million tonnes)	Imported quantities (million tonnes)	Yield per unit area (kg/ hm^2)
Soybean	1809.18	8851.28	1938.74
Corn	26,077.89	\	6316.70
Wheat	13,359.63	348.76	5630.40
Rice	23,588.40	254.57	12,961.50

Note Data in the table were from the National Bureau of Statistics

that in 2019, China needs 55 million hm^2 of arable land for soybeans, requiring 0.0387 hm^2/person; 41.2 million hm^2 of arable land for corn, 0.0293 hm^2/person; 24.3 million hm^2 of arable land for wheat, 0.0173 hm^2/person; and requiring 18.4 million hm^2 of arable land for rice, with a cultivated area of 0.0133 hm^2/person. The calculations showed that of the four major food crops, the largest area of arable land was required to grow the number of soybeans needed to meet the country's production needs and the largest area of arable land per capita.

According to the Chinese Third National Land Survey (referred to as 'the Third Survey'), Chinese arable land in 2019 covered an area of 128 million hm^2. Supposed Chinese soybean production is to reach complete self-sufficiency. In that case, 43% of the country's arable land is needed for soybean cultivation, which is difficult to achieve because other food crops also require arable land for production.

14.3 The Problems of Chinese Arable Land

The quantity and quality of arable land are related to national food security. The problem of arable land in China was mainly reflected in two aspects: The amount of arable land per capita was small, with only 0.0907 hm^2/person, far less than the world average of 0.32 hm^2/person. On the other hand, the quality of arable land was a serious problem. The problems of acidification, salinisation, and desertification of arable soils were prominent, and the phenomena of 'non-agricultural,' 'non-foodisation,' and 'abandonment of farmland' were severe.

14.3.1 Inadequate Amount of Arable Land Per Capita

Chinese arable land resources were characterised by low per capita occupancy, low-quality arable land, and low reserves of arable land (Shan and Wu 2018). According to the primary data of 'the Third Survey,' the arable land in China was 7.53 million hm^2 less than the amount of arable land in the Second National Land Survey (referred to as 'the Second Survey'). Based on a population of 1334.5 million in 2009, Chinese arable land per capita was about 0.1013 hm^2/person by the end of 'the Second Survey.' Based on the total population of 1410.08 million in 2019, by the end of 'the Third Survey,' the arable land per capita in China was 0.0107 hm^2/person less than that at the end of 'the Second Survey,' which was only 89% of 'the Second Survey.'

The reduction in the area of arable land has led to a reduction in the area sown for food and a decrease in the yield of food production, which has harmed the country's food security.

14.3.2 Soil Degradation on Arable Land is a Significant Problem

On 24 May 2021, China's Ministry of Ecology and Environment released the *2020 China Ecological and Environmental Status Bulletin* (*2020 China Ecological and Environmental Status Bulletin*), which shows the soil issues in Fig. 14.1.

Data from the Ministry of Agriculture and Rural Affairs' *2019 National Arable Land Quality Rating Bulletin* (*2019 National Arable Land Quality Rating Bulletin*) are shown in Table 14.2.

These series of soil degradation problems have severely restricted the use of arable land resources to the detriment of the country's food security and ecological safety.

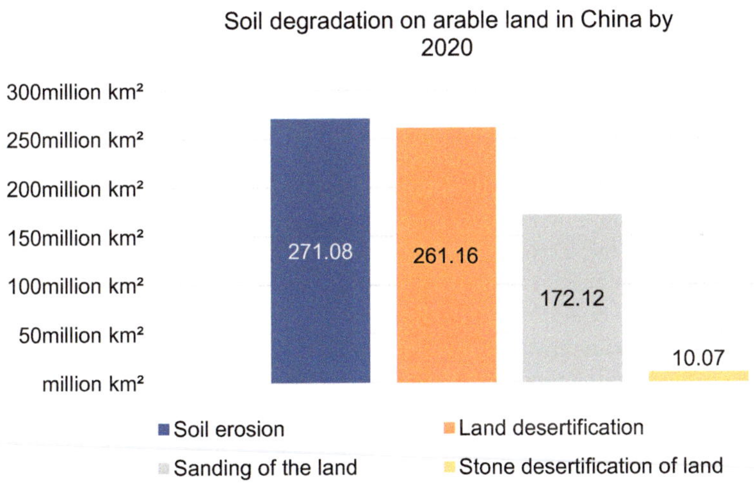

Fig. 14.1 Data on China's soil issues to 2020

Table 14.2 Data on soil issues in the *2019 National Arable Land Quality Rating Bulletin*

Regions	Soil issues	Area (hm^2)
Northeast China	Soil acidification, salinization, submergence	2.33 million
Inner Mongolia	Low soil nutrient content and severe soil erosion	4.27 million
Loess Plateau	Severe soil erosion and shallow soils	6.2 million
The middle and lower reaches of the Yangtze River	Nutrient-poor soils	4.6 million
The Yellow and Huaihai Sea region	Shallow soils	2.27 million
Southern China	Soil acidification, salinization	2.87 million
Southwest China	Soil acidification, submergence	4.53 million
Ganxin District	Salinization, desertification, sanding	1.8 million
Qinghai-Tibet region	Thin and nutrient-poor soils	0.7million

14.3.3 The 'Non-agricultural,' 'Non-foodisation,' and 'Abandonment of Farmland' in China Are Prominent Problems

According to the data of 'the Third Survey,' in the past ten years, there was a net loss of 7.47 million hm^2 of arable land to forest land and a net loss of 4.2 million hm^2 to garden land. China has implemented the 'rural revitalization strategy' to promote the revitalization of 'people, land and industry' and the organic integration of the 'Production-Life-Ecology' in the countryside (Niu et al. 2020). However, in the revitalization process, there were still problems of arable land abandonment (Zhang 2019) and 'non-foodisation' induced by unreasonable revitalization measures. Kong et al. (2020) concluded the 'non-foodisation' of arable land in China was 27%, with as much as 46% in southwest China. The total area sown for food crops declined from 80.36% in 1978 to 69.94% in 2019 (Kong 2021). The scale of abandonment in Chinese main grain-producing areas amounted to 4.05 million hm^2, mainly concentrated in northeastern Heilongjiang, southern Inner Mongolia, and northwestern Jilin, with a 'T'-shaped distribution (Li et al. 2021).

The phenomenon of 'non-agricultural,' 'non-foodisation,' and 'abandonment of farming' are becoming increasingly severe, resulting in a reduction in the arable land area, changes in rural land resources, and changes in the rural ecological environment, which are not conducive to ensuring national food security.

14.4 Shortcomings in Chinese Arable Land Protection Policy

To ensure that the quantity of arable land does not decrease, the quality of arable land does not deteriorate, and safeguard the ecological environment, China has introduced a series of relevant laws and implemented regulations to protect arable land. *The Land Management Law* of 1998 confirmed the protection of arable land as one of the basic state policies of China in the form of legislation. The document *Outline of the National General Land Use Plan (2006–2020)* stated that 'should keep the red line of 120 million hm^2 of arable land.' (*Outline of the National General Land Use Plan (2006–2020)*) In 2017, the State Council promulgated the *Opinions on Strengthening Arable Land Protection and Improving the Balance of Occupancy and Replenishment* to improve the problem of 'unequal quantity and quality' in occupying and replenishing the land. The newly revised *Land Management Law* and *the Implementing Regulations of the Land Management Law* promulgated in 2019 stated that the State shall implement a compensation system for the occupation of arable land. Such as 'as much as is occupied, as much as is reclaimed,' and shall not reduce the total amount of arable land and shall not lower the quality.

Trade-offs emerge between ecological and productive land. The construction of an ecological civilisation is one of the priorities of the 14th Five-Year Plan, but to a certain extent, ecological construction requires using arable land. By 2030, about 8 million hm^2 of arable land will be returned to forest, grass, berms, and wetlands (Zhang 2020; Qi et al. 2018), and how to balance ecological construction and arable land conservation will become a significant difficulty in arable land conservation. The current Chinese *Land Management Law* has no detailed solution to the conflict between the need to occupy arable land for ecological construction and the preservation of local arable land reserves.

14.5 Reflections and Recommendations

14.5.1 Technical

Integrated Area-wide Land Management Using Nature-based Solution (NbS). Using the NbS concept, which fully respects the laws of nature and is based on natural self-healing, certain microorganisms are used to promote material circulation, energy flow, and crop growth in the soil. At the same time, it enhances ecological carbon sinks, helps carbon neutrality, eases the pressure on food security caused by climate change, and improves the ecological environment. A complementary human role when NbS programmes are insufficient to meet the demand for arable land in the land use sector. The use of new ecological materials, equipment, techniques, and technologies for comprehensive land management in the whole area, placing ecology, quality, and quantity on an equal footing, optimising the spatial layout of the 'Production-Life-Ecology,' and changing the current situation of fragmentation of arable land, inefficient use of arable resources and degradation of the quality of the ecosystem. The city of Chishui in Guizhou Province combined the methods of traditional and ecological to develop ecological design of materials and craft in land remediation according to local conditions, so that green and ecology were integrated into the process of remediation (Xu et al. 2021).

Soil Degradation Solutions. ① Based on analysing the causes and mechanisms of the soil problems, and following the actual situation, deeply integrated industry, academia, and research to strengthen the control of soil degradation risks and propose corresponding remediation and treatment measures (He et al. 2020). For example, wind and sand control belts can be constructed in the northwest desert areas to reduce soil erosion to a certain extent (Ying and Zhang 2021). Slaked lime's application improves soil acidification in south China (Xie et al. 2021). Biotechnology improves soil salinity (Li and Zhao 2020; Wang et al. 2020b). ② Improve the quality environment of arable land. Use less or no chemical fertilizers and pesticides to control soil acidification, salinization, and soil pollution on arable land.

Increase Food Production. ① The restructuring of food cultivation positively affects the increase in food production (Zhu et al. 2013; Miao 2015). Implementing intercropping, interplant, and crop rotation can effectively balance soil nutrients, improve the structure of soil aggregates, increase soil fertility, make it less prone to succession barriers and guarantee food production. ② Agricultural mechanisation helps drive increased food production. Agricultural mechanisation helps improve agricultural production efficiency and mitigate the risk of declining food production due to labour shortages.

14.5.2 Policy

Establishing a Sound Ecological Compensation System for Arable Land. Arable land is an ecosystem; its construction and protection are essential to ecological construction. Establishing a proper ecological compensation system for arable land, emphasizing financial and technical compensation. Establish an ecological compensation fund for arable land, and divide the ecological compensation levels for arable land. Pay the corresponding ecological compensation to areas where arable land has been converted into ecological construction land due to ecological construction according to the divided levels. Technical compensation is mainly for implementing ecological agriculture without changing the area of arable land. Adhering to the viewpoint of 'teaching people how to fish,' teaching arable land owners advanced ecological agricultural cultivation techniques, developing ecological agriculture, and forming a virtuous circle between ecological construction and arable land protection. In addition, the compensation system for arable land protection, as a part of the solid ecological compensation system, government has also proposed to increase the compensation for arable land protection in the *Protection of Arable Land (Draft Law)*. Provinces in China such as Shaanxi Province and Shandong Province have introduced corresponding temporary measures to incentivise the protection of arable land.

Establishment of Institutions Related to the Transfer of Agricultural Land. Each province and municipality establishes an inspection body for the transfer of agricultural land and regularly reports the inspection results to the State Land Inspectorate. Land transfer can reduce the rate of 'abandonment of farmland.' (Guo et al. 2020) To ensure that the area of arable land and food production remains unchanged after the transfer of agricultural land, a detailed understanding of the inflowing party should be made before the transfer, including the use of the agricultural land after the transfer, the inflowing party's financial status and production and operation status. The inspectorates of agricultural land transfer in each province and city should supervise the whole process of agricultural land transfer and impose appropriate penalties on inflowing parties who do not use the land for its intended purpose, and on those who are involved in the transfer work who take bribes or other problems that make the transfer use different from the intended use. At the same time, report the inspection results to the land inspection bureau of the corresponding jurisdiction to curb the development of 'non-agriculture,' 'non-foodisation,' and 'abandonment of farming.'

Develop Criteria for Identifying the Quality of Degraded Arable Land. Each province and municipality sets the criteria for determining the damage to the cultivated layer according to the actual situation, and introduces corresponding penalties and regulations. Reference can be made to the requirements for determining damage to the cultivated layer developed by Hao et al. (2021). Divide the damage to the cultivated layer into three levels: undamaged, regular, and severe, and a 'cultivated layer damage system' is established accordingly. At the same time, link the credit rating to the degradation of arable land caused by human factors. The credit rating will be restricted according to the damage land's severity. Will order the person responsible for the damage to restore the damaged layer until it is restored to its original level before the bad credit is removed from the credit record.

Enhancing Education and Awareness of Multiple Actors in Arable Land Protection. The plurality of subjects here refers mainly to the masses, particularly the owners of arable land. Public education and awareness of arable land protection are essential to our basic national policy of 'effective protection of arable land.' Improve the incentive and guarantee mechanism to encourage people to participate in the protection of arable land. Propose new ideas for arable land protection, supervise the government's implementation of relevant policies, stimulate the public's awareness of the subject of arable land, and improve the level of arable land protection (Liu et al. 2020).

Acknowledgements This work was supported by the Ecological Evaluation of Terraced Fields in Ecological Isolation Zone of Plateau Red Earth Area, supported by Yunnan Provincial Department of Education Fund (Grant No. 2022J0300); and Trade-offs and Synergies of Ecosystem Services in Terraced Fields with Ecological Isolation Zones, supported by Yunnan Provincial Agricultural Joint Special Youth Project (Grant No. 2018FG001-092).

References

Anderson WB, Seager R, Baethgen W, Cane M, You L (2019) Synchronous crop failures and climate-forced production variability. Sci Adv 5(7):eaaw1976

Chen RS, Guo XN, Xiong B et al (2021) Climate change, land degradation and food insecurity: linkages and potential solutions. Acta Ecol Sin 41(07):2918–2929

China Agriculture Outlook Report (2021–2030). https://aocm.agri-outlook.cn/weixin/Public/2021/report/.

Cui QF, Pu MZ, Wang GG et al (2020) International grain export restrictions and China's grain security under the COVID-19 epidemic. Acad J Zhongzhou (04):20–26

Fan YT, Jin XB, Xiang XM et al (2018) Evaluation and spatial characteristics of arable land multifunction in southern Jiangsu. Resour Sci 40(5):980–992

Fu ZP (2021) India's food security crisis and its countermeasures under the continuous impact of COVID-19. South Asian Stud Quar (01):95–108+158

Guo BB, Fang YL, Zhou YK (2020) Influencing factors and spatial differentiation of cultivated land abandonment at the household scale. Resour Sci 42(4):696–709

Hao SH, Wu KN, Dong XR et al (2021) Identification criteria of cultivated horizon damage for "non-grain" cultivated land. Chin J Soil Sci 52(05):1028–1033

He Y, Liang XX, Pan RX et al (2020) The development process and prospect of soil pollution prevention and control in China. Chin Agric Sci Bull 36(28):99–105

Huang XM, Wang Q, Xiao T (2021) The food trade policy uncertainty and global governance under the impact of the COVID-19 pandemic. Intertrade 06:47–55

Kong XB (2020) Problems, causes and countermeasures of the "non-foodisation" of cropland. China Land 11:17–19

Kong XB (2021) Transition objectives, contents and ways of cultivated land protection in china under the background of rehabilitation system. J Soc Sci Hunan Normal Univ 50(03):1–12

Li LY, Zhao YL (2020) Discussion on the improvement measures of coastal Saline-alkali Land in Yingkou City. Territory Nat Resour Stud 04:71–73

Li YL, Ma WQ, Jiang GH et al (2021) The degree of cultivated land abandonment and its influence on grain yield in main grain producing areas of China. J Nat Resour 36(06):1439–1454

Liu LH, Zhang BX, Liu XH (2020) Compensation of provincial cultivated land protection in China from the dual perspectives of food security and ecological security. Trans Chin Soc Agric Eng (Trans CSAE). 36(19):252–263

Miao SS (2015) Sustainability of "ten consecutive increase" in core grain production area based on the perspective of grain planting structure adjustment. J Agro-Forest Econ Manage 14(6):567–576

Niu SD, Fang B, Cui C et al (2020) The spatial-temporal pattern and path of cultivated land use transition from the perspective of rural revitalization: Taking Huaihai Economic Zone as an example. J Nat Resour 35(08):1908–1925

Outline of the National General Land Use Plan (2006–2020) (www.gov.cn)

Qi XX, Zhang ZH, Huang XJ (2018) The contradiction of cultivated land protection in the new era and its innovative countermeasures. China Land Sci 32(08):9–15

Shan JM, Wu YZ (2018) A comparison of domestic and international arable land conservation and insights. Zhejiang Land Resour 7:21–24

State of Land and Water 2021 report (fao.org)

Su J, Ji R (2020) The supply and demand analysis on soybean and corn in China 2020. China Policy Rev 03:77–82

Wang D, Zhao YG, Ma R (2020a) Effects of microbial fertilizers on soil improvement and bacterial communities in saline-alkali soils of lycium barbarum. J Agric Biotechnol 28(08):1499–1510

Wang XJ, He YP, Jiang HP (2020b) China's food security during the 14th five-year plan period: situation problems and countermeasures. Reform 09:27–39

Wang XM, He W, Yang XW et al (2021) Impact of COVID-19 pandemic on global food security and countermeasures. J Agric Sci Technol 23(5):1–7

World Population Prospects projections report, 2019. https://www.un.org/development/desa/zh/news/population/word-population-prospects-2019.html.
Xie HY, Chen SR, Zhang Y et al (2021) Research progress on causes of soil acidification & acid soil improvement technology in South China. Hunan Agric Sci 2:104–107
Xu YL, Zhao YC, Zhai PM (2020) Advances in scientific understanding on climate change and food security from IPCC special report SRCCL. Climate Change Res 16(1):37–49
Xu XL, Pan BJ, Song JL (2021) major land remediation project in Wumeng mountain region of guizhou province exploring the effectiveness of construction-taking Chishui City as an example. J Green Sci Technol 23(20):189–192
Ying EY, Zhang LC (2021) Soil and water conservation and desertification control ideas. Tech Superv Water Resour (11):70–72+163
Zhang MJ (2019) An empirical study on the problem of cultivated land reclamation in the process of rural revitalization. Econ Res Guid 30:26–32
Zhang JX (2020) Research on cultivated land protection mechanism from the perspective of ecological civilization. Value Eng 39(18):80–81
Zhu J, Li TX, Lin DY (2013) Reflections after the "ninth consecutive increase": Analysis of the contribution of internal restructuring of food and future potential. Issues Agric Econ 34(11):36–43,110–111. 2019 National Arable Land quality rating bulletin. (moa.gov.cn). 2020 China ecological and environmental status bulletin. http://www.mee.gov.cn/hjzl/sthjzk/zghjzkgb/202105/P020210526572756184785

Chapter 15
Investigation Results and Countermeasures on the "Non-grainization" of Cultivated Land in Yunnan, Guangdong, and Guangxi

Xingmei Huang, Gengjie Zhang, Huiyong Hu, Honggang Zheng, and Shuxia Liu

Abstract Cultivated land is the basis for ensuring food security, but in recent years there has been a severe problem of "non-grainization" of cultivated land in China. We had research work in Yunnan, Guangxi, and Guangdong through field surveys, indoor discussions, and household questionnaires to investigate the situation and causes of eucalyptus planting on cultivated land. The results indicated that: (1) Eucalyptus was planted in large quantities in the 3 provinces. Eucalyptus planted area in Guangxi reached more than 2 million hm^2 in 2017, ranking first in the country. In 2019, more than 289,000 million hm^2 eucalyptus were planted in Yunnan and 1.72 million hm^2 in Guangdong. (2) Many people planted eucalyptus based on economic interests, and most farmers in Guangdong and Guangxi had little desire to convert plots planted

X. Huang
College of Resources and Environment, Yunnan Agricultural University, Kunming 650201, Yunnan, China

G. Zhang (✉) · H. Zheng · S. Liu
Land Resources Science and Technology Engineering Center, Yunnan Agricultural University, Kunming 650201, Yunnan, China
e-mail: jiegeng@ynau.edu.cn

Scientific and Technological Innovation Team for Optimizing the Land Allocation and Ecological Remediation in Southwest China's Diverse Regions, Ministry of Natural Resources, Kunming, China

Field Scientific Observation and Research Bases of Land Use in Yunnan Luliang, Ministry of Natural Resources, Kunming, China

Engineering Laboratory of Land Resources Utilization and Protection in Yunnan Province, Kunming, China

H. Hu
Undergraduate Academic Affairs Office, Yunnan Agricultural University, Kunming 650201, Yunnan, China

C. Yuan et al. (eds.), *Proceedings of 4th International Conference on Resources and Environmental Research—ICRER 2022*, Environmental Science and Engineering,
https://doi.org/10.1007/978-3-031-31808-5_15

with eucalyptus into cultivated land. (3) Eucalyptus plantations were more damaging to the soil, while man could improve the quality of the cultivated layer through engineering and biochemical measures. (4) There were fewer available test data and studies on the impact of eucalyptus planting on land. (5) Recreating existing plots planted with eucalyptus into cultivated land included root removal, land leveling, and soil fertilization. It analyzes why farmers plant eucalyptus from 4 aspects: policy, economy, law, and other reasons. Finally, it proposed countermeasures to safeguard the quantity and quality of cultivated land and maintain food security in China.

Keywords Non-grainization · Cultivated Land · Eucalyptus planting · Reasons · Countermeasures

15.1 Introduction

National security is an essential cornerstone for the country's stability, and food security is vital to the overall national security concept in the new era (Liu and Zhai 2021). To ensure national food security, China has implemented a strict policy of cultivated land protection and strictly remained above the redline of 12 million hm^2 of cultivated land (Cultivated land protection system). With the accelerated pace of agricultural restructuring (Chen and Sun 2017), the phenomenon of "non-grainizaition" utilization in distinct degrees has emerged in various places (Kong 2021; Yang et al. 2022), leading to the destruction in irrigation and drainage systems and road facilities of the original grain fields, the disruption of soil layers, a decline in fertility, and drastic changes in the intensity, depth, and type of use of cultivated land (Yao 2022). It has endangered the quantity and quality of cultivated land and has impacted national food security. During the "non-grainizaition" utilization of cultivated land, the phenomenon of occupying cultivated land for forest cultivation has been relatively in general. Eucalyptus has been particularly commonly planted because of its fast growth rate, a wide range of uses, and high economic benefits (Huang 2021; Oil-combined eucalyptus). According to statistics, by 2017, the planting area of eucalyptus in China was about 4.6 million hm^3, accounting for 6.5% of the Man-made forest area. The annual timber production was 30 million m^3, accounting for 26.9% of China's yearly timber production. Of the Chinese eucalyptus plantations, 80% were distributed in coastal provinces such as Guangxi, Guangdong, Hainan, and Fujian. Among them, the planting area of eucalyptus in Guangxi was about 1.87 million hm2, accounting for 41% of the national eucalyptus planting area, Guangdong 24%, Hainan 6%, and Fujian 11% desperately (Eucalyptus logs distribution). In addition, there was a certain amount of diffusion in Yunnan, Sichuan, Chongqing, and other provinces. Based on a thorough investigation and demonstration of the current situation of fast-growing timber forest plantations in China, the research team selected Yunnan, Guangxi, and Guangdong, where eucalyptus plantations covered a large area and occupied typical cultivated land for eucalyptus plantations. Through field surveys, indoor interviews, and household questionnaires, we discussed the destruction of the cultivated layer because of the "non-grainization" from cultivated land to eucalyptus plantation. By understanding the destruction on the tilth layer by planting fast-growing trees in dryland, it is expected that this work would support the government to issue the criteria for identifying the damaged state of tilth layer as the cultivated land treated as "non-grainizaition" land. Meanwhile, some theoretical evidence could be given to improve cultivated land protection policy for ensuring food security in China.

15.2 Study Area and Methods

Based on the data from the Third National Land Resource Survey (referred to as 'the Third Survey'), combined with research results and monitoring data from relevant departments of agriculture and rural areas, forestry and grass industry, and colleges and universities, we sorted out and clarified the scale and distribution characteristics of the types of fast-growing forest or timber trees grown in drylands in Yunnan, Guangdong and Guangxi provinces. We determined the study counties and sample points to carry out the field investigations, indoor discussions, and household questionnaires (Appendix 1).

(1) Selected three sample points in each survey county, and two plots were set up in each sample point: one typical plot (a large area of eucalyptus planted in the original dryland) and one comparative plot (the surrounding the specific plot with the same conditions for growing grain crops). Key indicators of cultivated land, such as the thickness of the tilth layer, soil configuration, organic matter content, soil pH, degree of weed invasion (such as noxious weeds, solid invasives), and pollution were recorded according to the eucalyptus plantation plots and surrounding grain plots (Appendix 2). The key indicators were measured in Table 15.1, and related images were shown in Fig. 15.1.
(2) Held indoor discussions to discuss the situation of the eucalyptus planting, the policy of the Grain to Green Program (GTGP), the utilization of cultivated land, and the impact of eucalyptus plantation on the tilth layer of land with relevant experts and leaders from the Department of Agriculture and Rural Affairs, the Forestry and Grassland Administration, the Territorial Planning and Design Institute, and the Land Ecological Restoration Centre.
(3) Carried out the household questionnaire survey with local farmers (Appendix 2). The designed questions mainly include "whether there is a plot of planting eucalyptus trees at home," "whether planting eucalyptus trees brings higher economic income compared with planting food crops," and "what are the attitudes towards plots of land planted with eucalyptus."

Table 15.1 Key indicator measurement methods

Key indicator	Measurement methods
Soil texture	Visual and hand measurement
Tilth layer thickness	On-site measurement
Organic matter content	The potassium dichromate volumetric method
Soil pH	The potentiometric method
Degree of weed invasion	On-site observed
bulk density	Circular cut method

Fig. 15.1 Section and landscape views. *Note* The first group was eucalyptus, and the second group was cultivated land

15.3 Results

According to data collection and collation, field research, indoor discussions, and household questionnaires, the following results were obtained regarding the occupation of cultivated land by eucalyptus planting and the destruction of the tilth layer.

15.3.1 Occupation of Cultivated Land for Eucalyptus Planting

The area of eucalyptus in Yunnan Province reached more than 2.89×10^5 hm^2, and the living wood-growing stock of eucalyptus came to 23 million m^3 in 2019. Among the 16 cities, Pu'er City had the largest planting area with 1.4×10^5 hm^2, accounting for 48.43% of the total eucalyptus planting area in Yunnan Province, followed by Chuxiong City, 5×10^4 hm^2, 17.30% of the total eucalyptus planting area in Yunnan Province. The minor distribution was Diqing prefecture, only 4.7 hm^2 in the whole region.

The data provided by the Yunnan Forestry and Grassland Bureau, in the new round of the GTGP from 2014 to 2019, restored 7.7×10^5 hm^2 of arbor forest land and planted a large proportion of the dryland with eucalyptus.

Because the blessed with natural conditions for the development of eucalyptus in Guangdong Province, since 2000, breakthroughs have been made in breeding new eucalyptus varieties and research on cultivation techniques, which has enhanced the confidence of the masses in afforestation to become rich and promoted the large-scale development of eucalyptus plantations. According to the 2019 resource data, the province had 17.23×10^5 hm^2 of eucalyptus, including 1.94×10^5 hm^2 of ecological public welfare forests and 15.29×10^5 hm^2 of commercial forests. The total accumulation volume of eucalyptus was 10^6 million m^3, accounting for 18% of the total forest accumulation volume in the province.

The area of eucalyptus in Guangxi Zhuang Autonomous Region reached more than 2×10^6 hm^2, accounting for nearly half of the total area of eucalyptus in the country; the living wood-growing stock of eucalyptus reached 150 million m^3, one-fifth of the total forest accumulation volume in the region in 2017, which ranked first in the country regarding the eucalyptus area, growth volume, and accumulation volume.

In the area of eucalyptus, state-owned forest farms account for 14.3%, and collective and individual 85.7%, of which 2.33×10^5 hm^2 were owned by enterprises and large households of more than 333.3 hm^2 in size. More than 90 species of eucalyptus were in the region, mainly distributed in the area south of the Tropic of Cancer. According to the Third Survey results, the GTGP restored 6.12×10^5 hm^2 of arbor forests and planted more than 90% with eucalyptus on dryland plots.

15.3.2 *Study Results*

(1) It was a fact that the "non-grainization" of cultivated land in some areas was because of the high economic benefits of planting eucalyptus.
(2) Soil problems such as declining soil organic matter and increased soil capacity thinning of the tilth layer from the planting of eucalyptus in Yunnan and Guangxi were more notable than in Guangdong because of the poor management and utilization.
(3) The quality of the tilth layer could be improved through engineering and biochemical measures to restore tillage and increase food production.
(4) There was less monitoring and research on the impact of eucalyptus planting on cultivated land. However, it was have been paid more attention to the impact of planting eucalyptus on soil, water resources, biology, and pest control.
(5) Over half of the farmers in the administrative regions of Guangxi and Guangdong supported the change of cultivated land planted with eucalyptus into economic fruit and ecological forests, but they did lack of willingness to transform it into cultivated land, while farmers in Yunnan province were slightly more willing to support.
(6) The reclamation of land planted with eucalyptus to cultivated land mainly focused on three aspects: root removal, land leveling, and soil fertilization, the cost involved was related to the local economic situation, and the time of reclamation was unrestricted.

15.4 Cause Analysis

15.4.1 *Policy Reasons*

The Grain to Green Program. National had the requirement of expanding the scale of the GTGP (www.gov.cn) to achieve the goal of returning all cultivated land on slopes above 25° and returning all cultivated land between 15 and 25° as far as possible. Yunnan Province has planned to return 10.62×10^5 hm^2 in a new round of the GTGP from 2014–2020. According to the data from Yunnan Forestry and Grassland Bureau, 7.7×10^5 hm^2 of the new round of the GTGP planned for the province from 2014–2019 ranked second in the country.

Afforestation subsidy. To ensure the smooth implementation of the task of the GTGP, the government implemented a series of afforestation subsidies (Afforestation subsidy) to mobilize farmers' enthusiasm for planting trees. Eucalyptus was gradually becoming a new addition to the forestry industry because of the characteristics of wide adaptability, resistance to drought and barrenness, easy planting (Huang 2021), and remarkable economic and ecological benefits (Huang 2021; Oil-combined eucalyptus). For example, No. 22 〔2002〕 of the People's Government of Guangxi Zhuang Autonomous Region provided preferential policies, such as halving the collection of afforestation funds for creating fast-growing forests and providing financial discounts on loans for afforestation, and gave priority to timber harvesting indicators, and other preferential policies. The forestry department stipulated that the critical project afforestation took, such as crucial shelter forests, prioritized the development of eucalyptus trees and enjoyed the afforestation subsidies, which have dramatically mobilized the enthusiasm of the whole society to plant eucalyptus trees.

15.4.2 Economic Reasons

Export of eucalyptus oil. Eucalyptus had many uses and high economic value (Oil-combined eucalyptus). The areas of eucalyptus planting have been increased by farmers spontaneously because the price of eucalyptus oil has continued to rise, coupled with the national policy of banning logging in natural forests (12. Thirteenth Five-Year Plan; Fu 2016), and the characteristics of fast-growing, easy-to-manage, and high benefits.

Development of the timber industry. With the development of the economy, the annual consumption of timber in China has been rose from 214 million m^3 in 2002 (Zhang 2003) to 610 million m^3 in 2016 (Chen 2016), of which the dependence on imported timber exceeded 50% and was growing year by year. For example, Guangxi Zhuang Autonomous Region vigorously introduced wood-based panel processing technology. By 2016, the number of wood processing enterprises with eucalyptus wood as the primary raw material has grown to 16,000, including more than 1,200 wood-based panel manufacturers. Wood-based panels' output reached more than 36 million m^3, ranked 3 in the country. The medium-density fiberboard and particleboard production were more than 10 million m^3, ranked 1 in the country. The massive demand for wood-based panels has driven the rapid development of the eucalyptus planting industry.

Comparative economic advantages. Due to the low benefit of traditionally planting food crops (Xu et al. 2021) and suitable climatic conditions for the forestry and fruit industry, resulting agricultural restructuring on cultivated land in some areas (Kong 2020). For example, in the verification of reserve and supplementary cultivated land have been carried out by the Ministry of Natural Resources of Guangxi Zhuang Autonomous Region in 2019, most of the problem plots with unrealistic additional cultivated land were commercial crops or fruit trees. Suppose the farmers wanted to continue using the remaining supplemental cultivated land indicators, they must be changed commercial crops or fruit trees to food crops, and the subsidy for changing tangerine in Guinan reached 60–80,000 RMB per 0.07 hm^2, while other areas also needed 20–50,000 RMB per 0.07 hm^2. The challenge in localities and impoverished regions was implementing rectification funds. At the same time, the low economic benefits of grain crops and farmers' reluctance to use the land for growing grain crops made it more challenging to complete the rectification work of reserve and supplementary cultivated land.

15.4.3 Legal Reasons

(1) Although Chap. 4 of the current *The Law of Land Administration of the People's Republic of China*, entitled "Protection of Cultivated Land," proposed to "strictly control the conversion of cultivated land into non-cultivated land," it only emphasized the occupation of cultivated land by non-agricultural construction and did not explicitly address the situation of the conversion of cultivated land into agricultural land. At the same time, the enforcement of the transformation of cultivated land into other grounds was poor, and there was a lack of punishment laws for destroying the tilth layer. The communication with grassroots cadres confirmed this problem during the research.
(2) The current *Regulations on the Protection of Basic Farmland*, Article 17, paragraph 2, "prohibits any unit or individual from occupying basic farmland to develop forestry and fruit industries and digging ponds for fish farming," but there were no supportive measures.

15.4.4 Other Reasons

(1) Due to the less workforce and the remaining labor force in rural areas, those who were old, weak, ill, or disabled, while the economic benefit of planting eucalyptus was higher than that of grain planting, and the labor cost was low, which no more labor was needed after the first year of fertilization. On the other hand, families with no or weak labor force guaranteed the right to occupation cultivated land, which avoided encroachment by neighbors through planting eucalyptus.
(2) Influenced by the affirmation of woodland rights, some farmers chose to plant fast-growing eucalyptus on cultivated land to change the land contract rights from 30 to 70 years through land use changes. Farmers were expected to be compensated through the eucalyptus trees attached to the ground for the areas near the project construction.

15.5 Conclusion and Countermeasures

The high benefits of eucalyptus cultivation, the low cost of labor, and the policy have encouraged farmers to grow eucalyptus on a large scale; the lack of legal provisions has exacerbated the phenomenon of "non-grainization" of cultivated land; eucalyptus cultivation has led to problems such as soil consolidation and a decline in soil organic matter, which has had a negative impact on the local environment, national food security and social stability. Food is a matter of livelihood, and food security is an important foundation for national security. Therefore, this paper argues that specific measures should be taken in four areas: policy, economic, legal, and others to guarantee national food security.

15.5.1 Policy

(1) Suggest that the policy of the GTGP is implemented in zones and not be a uniform standard. Appropriate policy preference in areas with weak conditions and large returning farmland to forests. Retention of some eucalyptus forests as ecological land to prevent wind and sand fixation and improve the ecological environment in ecologically fragile areas.
(2) It is recommended that plots of land planted with eucalyptus are used as a reserve resource for supplemental cultivated land. The newly added cultivated land can be used as an additional cultivated land indicator for the balance of occupation to stimulate local enthusiasm for clearing eucalyptus and reduce the economic burden on local governments and local farmers.
(3) Improve the policy of household contract management and strictly implement "whoever contracts is responsible" to ensure that the use of cultivated land for planting food will not change and will not degrade the quality.

15.5.2 Economy

(1) The government can purchase grain from growers at a high price and then sell it to consumers at market parity to ensure enthusiasm for local food planting in some areas where economic conditions allow.
(2) Families with low labor are encouraged to subcontract their cultivated land to grain growers by policies or economic incentives and introducing advanced agricultural concepts and techniques to guide farmers to try intercropping.
(3) Guide the formation of local agricultural industries with unique characteristics (Li 2016) to improve farmers' income and enhance professional pride.

15.5.3 Law

(1) The situation when converting cultivated land into other agricultural lands for the development of the forestry and fruit industry is added to *The Law of Land Administration of the People's Republic of China*.

(2) Supplement the supportive measures for the occupation of bare farmland to develop the forestry and fruit industry in the *Protection of Basic Farmland Regulations.*
(3) There is a lack of punishment laws for destroying the tilth layer, and we suggest introducing legal provisions. For example: In the revision of the *Regulations on the Protection of Basic Farmland*, it is added: the penalty clause for the destruction of the tilth layer and the undamaged tilth layer is recommended to be included in the category of cultivated land; revegetation of cultivated layers with considerable damage; for the completely destroyed tilth layer, file a case and punish.

15.5.4 Other

(1) It is suggested that the standards for determining the damage to tilth layer be unified at the national level (Li et al. 2021). Based on the results of the survey indicated that divided the identification of "non-grainizaition" tilth layer damage into three types:

    Plots, where the tilth layer is not destroyed can continue to grow food crops without engineering measures or reduce the tilth layer's thickness by less than 5 cm, such as those planted with citrus, passion fruit, dragon fruit, and other vine or shrub fruit tree plots.

    More damaged tilth layer. That is, the thickness of the tilth layer is not up to standard, the depth of digging damage is between 5 and 10 cm, the root system of the original crop is well developed, and after restoration through engineering measures, food crops can continue to be grown. Such as planted trees, trees higher than 2 m, and planted eucalyptus and other fast-growing forests, resulting in a substandard thickness of the tilth layer, rigid and compact topsoil, and soil too acidic alkaline plots.

    Completely destroy the tilth layer, the depth of digging damage is more than 10 cm, it is no longer possible to continue to grow food crops, and the tilth layer is completely stripped or hardened, including mainly due to digging ponds for fish farming, development of a facility for agriculture and rural tourism, resulting in the destruction of the cultivation horizon of the plot.
(2) The studies on the impact of eucalyptus on soil (Zhang et al. 2022; Zhao et al. 2021), water resources (Yang et al. 2018), biology (Huang et al. 2022; Li et al. 2019), and pest control (Fu 2021) have focused on forest land. There is a lack of data on monitoring the land impact of planting eucalyptus on cultivated land. It is suggested to strengthen monitoring and research on the effect of planting eucalyptus on cultivated land.

Cultivated land is essential for food production (Kong et al. 2022). To solve the problem of feeding a population of 1.4 billion people, in addition to considering grain seeds (Wan 2022) and planting techniques (Qian 2018), one must also hold on to the foundation of the cultivated land. The Notice of the General Office of the State Council on Resolutely Stopping the "Non-agriculturalization of Cultivated Land states that those who illegally occupy cultivated land and permanent bare farmland for afforestation shall not verify the afforestation area and do not enjoy the financial subsidies policy. Plain areas should rationally formulate ecological construction goals such as greening and afforestation according to resource endowments. Must strictly control the GTGP within the scale and scope approved by the state. All the plots of land should be managed as shown on the map to immediately stop the greening and reforestation of cultivated land being illegally occupied.

Acknowledgements This work is supported by

(1) Ecological Evaluation of Terraced Fields in Ecological Isolation Zone of Plateau Red Earth Area, supported by Yunnan Provincial Department of Education Fund (Grant No. 2022J0300).

(2) Trade-offs and Synergies of Ecosystem Services in Terraced Fields with Ecological Isolation Zones, supported by Yunnan Provincial Agricultural Joint Special Youth Project (Grant No. 2018FG001-092).

Appendix 1

Household questionnaires form for fast-growing forest trees (eucalyptus) planting

Survey Date:

Survey contents		1. Does your fa-mily have a pl-ot of land for planting eucaly-ptus trees?		2. Does planting eucalypt-us trees bring higher econ-omic income than plantin-g food crops?			3. If considering land use change, what is your attitude towards the plots of land planted with eucalyptus trees?				Signature of investigat-or
		A. Yes	B. No	A. Yes	B. No	C. Do-n't un-dersta-nd	A. Prese-rvation of euca-lyptus t-rees	B. Con-versio-n to c-ultivated lan-d	C. Conv-ersion to a brokera-ge fruit forest	D. Con-version to the ecologi-cal for-est	
Investigato-rs' opinions	Investiga-tor 1										
	Investiga-tor 2										
	Investiga-tor 3										
	Investiga-tor 4										
	Investiga-tor 5										
	Investiga-tor 6										
	Investiga-tor 7										
	Investiga-tor 8										

(continued)

(continued)

Survey contents		1. Does your fa-mily have a pl-ot of land for planting eucaly-ptus trees?		2. Does planting eucalypt-us trees bring higher econ-omic income than plantin-g food crops?			3. If considering land use change, what is your attitude towards the plots of land planted with eucalyptus trees?				Signature of investigat-or
		A. Yes	B. No	A. Yes	B. No	C. Do-n't un-dersta-nd	A. Prese-rvation of euca-lyptus t-rees	B. Con-versio-n to c-ultivated lan-d	C. Conv-ersion to a brokera-ge fruit forest	D. Con-version to the ecologi-cal for-est	
	Investiga-tor 9										
	Investiga-tor 10										

Appendix 2

Basic information sheet on eucalyptus planting in Guangxi

Guangxi (Qinbei District, Qinzhou)							
		Yonglong Village, Pingji Town		Naliao Village, Xiaodong Town		Dadong Village, Dadong Town	
		Typical plot (eucalyptus)	Surrounding comparison plot (Grain)	Typical plot (eucalyptus)	Surrounding comparison plot (Grain)	Typical plot (eucalyptus)	Surrounding comparison plot (Grain)
Basic information	Longitude	22°12′49″ N	22°12′ 49″ N	22° 14′ 50″ N	22° 14′ 50″ N	22° 14′ 50″ N	22° 14′ 50″ N
	Latitude	108° 51′ 00″ E	108° 51′ 00″ E	108° 38′ 44″ E	108° 38′ 44″ E	108° 38′ 44″ E	108° 38′ 44″ E
	Altitude (m)	12	12	8	8	8	8
	Planting time (Years)	5	/	8	/	8	/
	Growing crops	Eucalyptus	Peanut	Eucalyptus	Rice	Eucalyptus	Peanut
	Planting density (Plant/ 0.066 7hm^2)	110	/	110	/	110	/
Key indicators	Soil texture	Loamy soil	Loamy soil	Sandy loam	Sandy loam	Sandy loam	Sandy loam
	Tilth layer thickness (cm)	21	41	25	28	14	20
	Organic matter content(g/ kg)	8.3	20.6	7.2	22.7	9.2	26.7
	Soil pH	4.26	5.33	4.31	5.42	4.11	5.52
	Degree of weed invasion	/	/	/	/	/	/
	Whether there is pollution	/	/	/	/	/	/
	bulk density (kg/cm^3)	1.42	1.31	1.45	1.32	1.42	1.3
Environmental conditions	Access to the road	Basically sound	Basically sound	Sound	Sound	Sound	Sound
	Irrigation and drainage facilities	Unsound	Unsound	Unsound	Unsound	Unsound	Unsound
Management measures	Seedling input	170	315	170	315	170	315

(continued)

(continued)

Guangxi (Qinbei District, Qinzhou)							
		Yonglong Village, Pingji Town		Naliao Village, Xiaodong Town		Dadong Village, Dadong Town	
		Typical plot (eucalyptus)	Surrounding comparison plot (Grain)	Typical plot (eucalyptus)	Surrounding comparison plot (Grain)	Typical plot (eucalyptus)	Surrounding comparison plot (Grain)
	Fertilizer and pesticide input	195	330	195	330	200	330
	Tilling frequency (Year/ Times)	0.5	2	/	2	/	2
	Economic income (Yuan/ 0.066 7hm^2)	2400	1600	2400	1600	2400	1600

Basic information sheet on eucalyptus planting in Guangdong

Guangdong (Leizhou City, Zhanjiang)							
		Fenyong High-tech Zone Huaqiao 16 team		Nashuang Village, Longmen Town		Longmen Town Happy 11 Team	
		Typical plot (eucalyptus)	Surrounding comparison plot (Grain)	Typical plot (eucalyptus)	Surrounding comparison plot (Grain)	Typical plot (eucalyptus)	Surrounding comparison plot (Grain)
Basic information	Longitude	21° 03′ 06″ N	21° 03′ 06″ N	21° 08′ 47″ N	21° 08′ 47″ N	21° 04′ 13.8″ N	21° 04′ 13.8″ N
	Latitude	110° 04′ 20″ E	110° 04′ 20″ E	110° 39′ 51″ E	110° 39′ 51 ″E	110° 04′ 09″ E	110° 04′ 09″ E
	Altitude (m)	23	23	31	31	82	82
	Planting time (Years)	5	/	6	/	1	/
	Growing crops	Eucalyptus	Chillies and maize	Eucalyptus	Sisal	Eucalyptus	Sugar cane, sisal and pineapple
	Planting density (Plant/ 0.066 7hm^2)	117	/	110	/	105	/
Key indicators	Soil texture	Sandy loam	Sandy loam	Sandy loam	Sandy loam	Sandy loam	Sandy loam
	Tilth layer thickness (cm)	28	31	20	31	19	30

(continued)

(continued)

Guangdong (Leizhou City, Zhanjiang)							
		Fenyong High-tech Zone Huaqiao 16 team		Nashuang Village, Longmen Town		Longmen Town Happy 11 Team	
		Typical plot (eucalyptus)	Surrounding comparison plot (Grain)	Typical plot (eucalyptus)	Surrounding comparison plot (Grain)	Typical plot (eucalyptus)	Surrounding comparison plot (Grain)
	Organic matter content(g/kg)	9.3–29.0	6.1–22.5	4.9–10.7	2.9–20.3	8.8–20.2	9.6–23.4
	Soil pH	4.29	4.77	4.15	5.36	4.7	4.99
	Degree of weed invasion	/	/	Ant	Ant	/	/
	Whether there is pollution	/	/	/	/	/	/
	bulk density (kg/cm^3)	1.4	1.2	1.4	1.1	1.2	1
Environmental conditions	Access to the road	Sound	Sound	Sound	Sound	Sound	Sound
	Irrigation and drainage facilities	Basically sound	Basically sound	Basically sound	Basically sound	Unsound	Unsound
Management measures	Seedling input	160	450	160	450	160	450
	Fertilizer and pesticide input	/	60	/	60	/	60
	Tilling frequency (Year/Times)	0.2	0.5	0.2	0.5	0.2	0.5
	Economic income (Yuan/0.0667hm^2)	3600	1600	4000	1900	4200	2800

Basic information sheet on eucalyptus planting in Yunnan

Yunnan (Mojiang Hani Autonomous County, Pu'er City)							
		Tongguan Town Dazhai Village Committee group		Xinwu Community, Tongguan Town		Kuaifa Village, Lianzhu Town	
		Typical plot (eucalyptus)	Surrounding comparison plot (Grain)	Typical plot (eucalyptus)	Surrounding comparison plot (Grain)	Typical plot (eucalyptus)	Surrounding comparison plot (Grain)
Basic information	Longitude	23°16′43″N	23°16′43″N	23°16′31″N	23°16′31″N	23°23′33″N	23°23′33″N
	Latitude	101°25′3″E	101°25′3″E	101°21′55″E	101°21′55″E	101°40′60″E	101°40′60″E
	Altitude (m)	1873	1873	1866	1866	1334	1334
	Planting time (Years)	5	/	5	/	6	/
	Growing crops	Eucalyptus	Grain	Eucalyptus	Grain	Eucalyptus	Grain
	Planting density (Plant/ 0.066 7hm^2)	120	/	120	/	120	/
Key indicators	Soil texture	Sandy loam	Sandy loam	Sandy loam	Sandy loam	Sandy loam	Sandy loam
	Tilth layer thickness (cm)	30	15	22	15	27	15
	Organic matter content(g/ kg)	20.1	23.7	20.6	22.6	23.4	25.9
	Soil pH	4.96	5.52	4.86	5.14	5.4	5.71
	Degree of weed invasion	/	Ant	/	Ant	/	Ant
	Whether there is pollution	Yes	No	Yes	/	Yes	/
	bulk density (kg/cm^3)	1.39	1.35	1.31	1.25	1.74	1.63
Environmental conditions	Access to the road	Sound	Sound	Sound	Sound	Sound	Sound
	Irrigation and drainage facilities	Unsound	Unsound	Unsound	Unsound	Unsound	Unsound
Management measures	Seedling input	70	120	80	125	75	120
	Fertilizer and pesticide input	130	220	140	230	150	260

(continued)

(continued)

Yunnan (Mojiang Hani Autonomous County, Pu'er City)							
		Tongguan Town Dazhai Village Committee group		Xinwu Community, Tongguan Town		Kuaifa Village, Lianzhu Town	
		Typical plot (eucalyptus)	Surrounding comparison plot (Grain)	Typical plot (eucalyptus)	Surrounding comparison plot (Grain)	Typical plot (eucalyptus)	Surrounding comparison plot (Grain)
	Tilling frequency (Year/Times)	/	2	/	2	/	2
	Economic income (Yuan/0.066 7hm^2)	7000	2100	7000	2200	6500	2000

References

Afforestation subsidy www.gov.cn

Cultivated land protection system. mnr.gov.cn

Chen QN, Sun NB (2017) Non-grain fertilization of farmland in Ningxia from the perspective of food security. Guangdong Agric Sci 44:154–159

Chen SH (2017) Analysis of China's wood consumption and resources in 2016. China For Ind 34–35

Eucalyptus logs distribution. chinabaogao.com

Fu MH (2016) Current situation and countermeasures of collective natural forest resources protection in Guangde county. Anhui Agricul Sci Bull 100–102

Fu BJ (2021) Eucalyptus silviculture technology and pest control technology. Contemp Hortic 44:69–70

Grain to Green Program. www.gov.cn

Huang S (2021) Technical analysis of eucalyptus stand assets valuation. Nongcun Shiyong Jishu 119–120

Huang C, Peng XR, Huang JR et al (2022) Exploration of the effect of artificial nurturing methods on the biodiversity of Eucalyptus plantation forests. J Green Sci Technol 24:35–38

Kong XB (2021) Transition objectives, contents, and ways of cultivated land protection in China under the background of a rehabilitation system. J Soc Sci Hunan Normal Univ 50:1–12

Kong XB (2020) Problems, causes, and countermeasures of "non-grainization" of cultivated Land. China Land 17–19

Kong XB, Chen WG, Wen LY (2022) Building the foundation of China's grain security with three security of cultivated land resources. Agric Econ Manage 1–12

Li JC (2016) Developing local special grape industry and promoting agricultural transformation and upgrading. In: Proceedings of the sixth annual academic conference of yunnan provincial science association and red river basin development forum——Topic I: transformation and upgrading of special industries in the red River Basin, pp 210–213

Li JY, Lin JY, Pei CY et al (2019) Eucalyptus plantations alter the diversity and structure of the soil arbuscular mycorrhizal fungal community. Acta Ecol Sin 39:2723–2731

Li C, Wang W, Li WC (2021) Impacts of "non- grainization" use on cultivated land quality. China Land 17–19

Liu XH, Zhai W (2021) National food security governance under the overall national security concept. Wisdom China, pp 30–33

Oil-combined eucalyptus. <ynls.gov.cn>

Qian ZW (2018) Outline Xi Jinping's strategic thought of national food security in the new era. J Hefei Univ Technol (Soc Sci) 32:29–35

Thirteenth Five-Year Plan. <cnr.cn>

Wan YX (2022) On the status and ways of crop seeds in ensuring national food security. Gansu Agric 03:68–72

Xu XY, Bao JC, Wang QC (2021) An analysis of the management of the phenomenon of "non-foodization" of cultivated land: the example of Lu Liang County, Yunnan Province. China Land 45–47

Yang JX, Chen LX, Li WT et al (2018) Exploring the impact of eucalyptus planting in reservoir areas on water conservation and soil erosion based on remote sensing technology. South China Agric 12:67–68

Yang HX, Jing YT, Zhu Y (2022) The current situation is the causes and countermeasures of "non-grainization of arable land in China. J Anhui Agric Univ (Soc Sci Edn) 31:33–37+112

Yao XJ (2022) Problems and countermeasures of "non-agriculturalization" and "non-grainization" of cultivated land. Agric Eng Technol 42:94–95

Zhang K (2003) China timber market research. For Econ 48–49

Zhang Z, An B, Fu J et al (2022) Soil fertility quality evaluation in eucalyptus plantation of state-owned Paiyangshan Forest Farm in Guangxi. Southwest China J Agricul Sci 1–8

Zhao QQ, Zhao XQ, Huang P et al (2021) Review of variation of soil physical and chemical properties among eucalyptus plantations of different ages in China. Eucalypt Sci Technol 38:47–52

Chapter 16
Urban Regeneration and Integration of Public Spaces of the Tumilaca River in the Department of Moquegua

Esenarro Doris, Ayala Yannet, Puchoc Lourdes, Ramirez Maria, and Zelada Maria

Abstract The importance of public spaces nowadays has become more important, so this research aims to propose a design that allows urban regeneration and therefore social integration for a better quality of life for the inhabitants of the Tumilaca River, one of the three tributaries of the Moquegua River basin, since there is a long history of disasters, whose occurrence is related to numerous natural events and climatic phenomena and changes. As a methodology, design strategies were applied with a sustainable approach, considering bioclimatic criteria appropriate to the environment, and as a result, a central plaza, sports areas, bicycle paths, trails and a viewpoint were implemented within the public space of the river. In conclusion, the study area has climatic characteristics that do not favor an adequate urban-environmental development for its intervention, and the use of the proposed clean energies allows mitigating environmental contamination, allowing taking advantage of the climatology of the place.

Keywords Urban regeneration · Integration of public spaces · Bioclimatic strategies

16.1 Introduction

Urban planning, a complex subject (Adhikari and Li 2013), gives us a vision of future procedures, in order to anticipate and propose solutions to potential problems (Steiner 2018), being a highly political activity, aimed at improving the quality of decision making, with the objective of serving the public interest (Levy 2016).

E. Doris (✉) · A. Yannet · P. Lourdes · R. Maria · Z. Maria
University Ricardo Palma, Lima, Peru
e-mail: doris.esenarro@urp.edu.pe

E. Doris
Specialized Institute for Research on Ecosystems and Natural Resources—INERN, Lima, Peru

C. Yuan et al. (eds.), *Proceedings of 4th International Conference on Resources and Environmental Research—ICRER 2022*, Environmental Science and Engineering,
https://doi.org/10.1007/978-3-031-31808-5_16

Initially concerned with an orderly arrangement of public spaces, urban planning now also focuses on environmental imperatives, not only to create new environmentally considerate spaces (Fitzgerald 2010), but also to anticipate potential environmental risks (Dickson et al. 2012) with adaptive and resilient approaches (Ganis 2015). One of the fundamental considerations of this project, is argued in the use and control of urban space in terms of green areas and recreation within the city, to give added value to deteriorated or inactive areas (Danjaji and Ariffin 2017) as the absence of infrastructure of the services provided by these systems such as recreation and leisure. The assembly of these are essential factors for the valuation of land use.

The Tumilaca River is one of the main tributaries that form the Moquegua River, joining at the heights of the city, surrounded by valleys and flowing into the Pacific Ocean. The wet basin covers approximately 225 km^2, has an average annual flow between January and March of 1067 m^3/s and at 75% persistence it is 0.683 m^3/s (https://www.midagri.gob.pe/), activating the dry creeks, resulting in overflows and floods in agricultural and urban areas. The urban morphology that has developed in the department of Moquegua represents the urban reality that afflicts the population in the area around the Tumilaca River. Throughout history, it has been hit by the severity of climatic phenomena and changes with unusual intensity (Estudios and y Prevención de Desastres 2003), due to the physical characteristics of the city, as well as causes in which the inhabitants, in search of spaces that shelter them, exceeded the limit, locating themselves in areas very close to rivers and streams, without taking into account the characteristics of the land and its surroundings. (Kuon and Cabello 2015) Nowadays, strategies and methodologies have been created in which attention to the maintenance and protection of riparian structures are applied to mitigate vulnerability to natural disasters or climatological phenomena (Hernández-Uribe et al. 2017), as part of the contribution to economic growth, due to the link to one of the main activities that governs the district, such as agriculture and the environment of its varied valleys (Vivienda and Construcción y Saneamiento 2016). However, there is little attention to the development of tourist activities, lack of recreational areas and lack of interest in public spaces. Therefore, the present proposal has as its objective the urban intervention and integration of public spaces in the Tumilaca Riverbank, improving the quality of life of its inhabitants, as well as the presence of infrastructure for these services, which will provide systems such as recreation and leisure, essential factors for the valuation of land use, as well as the impulse in the economic sector through the activation of the tourism sector.

16.1.1 Literature Review

Bioclimatic Strategies. These strategies benefit climatic conditions in the context of riverbank regeneration (Cengiz 2013), which strengthens the case for sustainable urbanism. Waterside planning can provide higher quality urban open spaces for pedestrians and cyclists (Pandit et al. 2020). Urban recreational functions, such as

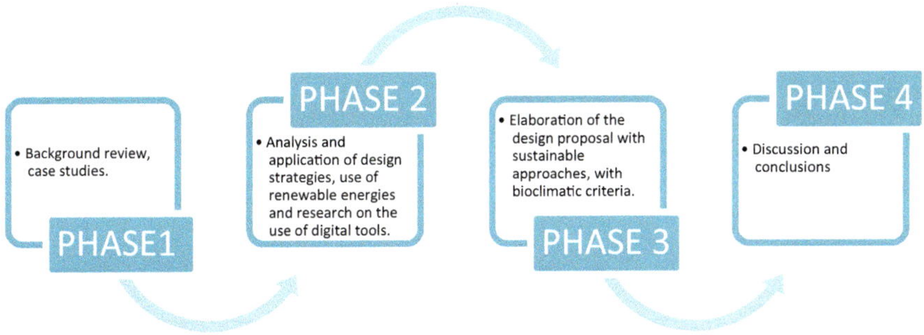

Fig. 16.1 Phase study

public parks, sports facilities, and playgrounds, are often located near riverbanks, which attracts riparian greenway projects (Balsas 2021).

Malecón Ribereño. The thematic riverside boardwalk will be a recreational area, which will house both the inhabitants and the tourist sector, where social interaction will be encouraged, being the visual perception of this project, the valley and the Tumilaca River basin, will integrate the environment with the activities that will be generated in the architectural monument (Portner 1992).

16.2 Methodoly

16.2.1 Methodological Scheme

Once the topic has been defined, as a result of the information in situ, the central question of the research is born and with it the objectives and variables that allow the hypothesis to be generated. In order to elaborate the research diagnosis and arrive at a result, it is necessary to have knowledge of the subject, the product of a theoretical framework backed by bibliographic support, as well as techniques and instruments for the collection and subsequent analysis of both general and specific information, which, once synthesized, will allow conclusions to be drawn that favor decision making for the development of an architectural proposal (Alberto and Aquino 2016).

Figure 16.1 shows the analysis of the field survey for the preparation of the report with a compilation of information from matrices to make a proposal.

16.2.2 Place of Study

This area is located in the Moquegua region, which is close to the Tumilaca River. It is often affected by the El Niño phenomenon, causing the river to overflow and

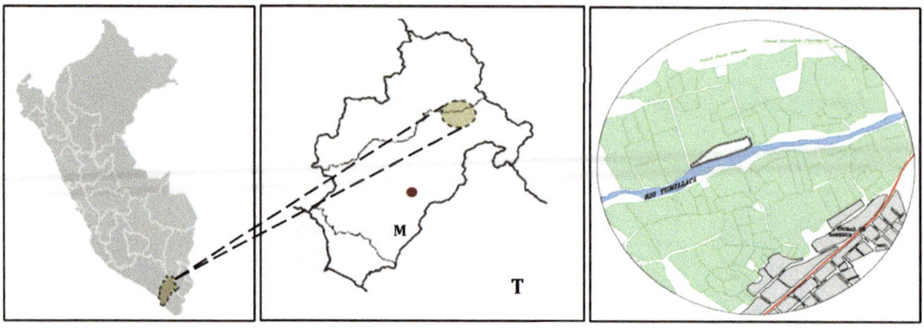

Fig. 16.2 Location of the study area

threatening the inhabitants living near the river. The study area is composed of city, valley and desert.

Figure 16.2 shows the intervention site and the location of the study area in the Department of Moquegua, Lima, Peru.

16.2.3 Climate Analysis

Solar Radiation

In Fig. 16.3, the following graphs show the brightest period of the year lasts 3 months from October to December, in which an average solar radiation of 6.8 km h/m^2 is maintained. The months with the highest radiation are in the summer and spring seasons, while the autumn and winter seasons have low levels.

Temperature and Precipitation. Moquegua is largely within the comfort zone; its climate is moderate with relative humidity and a wide range of temperatures throughout the year. The local population is comfortable from early morning to midday. The temperature remains relatively constant throughout most of the year; nights are cool with temperatures between 9° and 10°. Precipitation is scarce in most months of the year (April and November), however, there is an increase in the

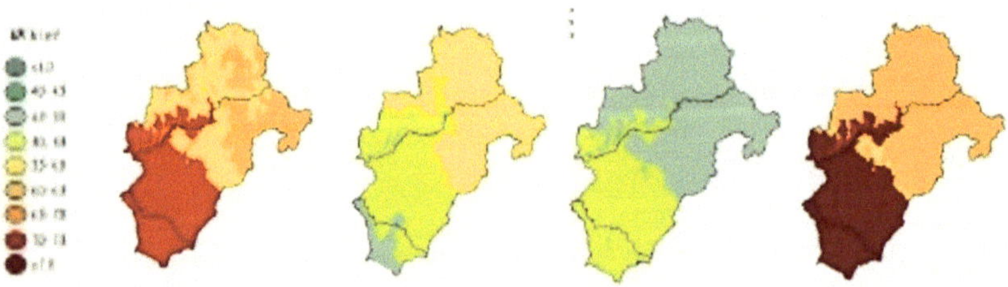

Fig. 16.3 Solar radiation

summer months with 8.1 mm. In the sector between 2500 and 5000 masl, precipitation increases with altitude, from approximately 150 mm in the lowest part to more than 500 mm in the Suni; precipitation in the Puna and Suni region (>3800 masl) is rain, hail, snow, drizzle and dew; occasionally there are violent storms accompanied by strong winds (speeds greater than 42 km/h).

Figure 16.4a, shows a stable temperature throughout the year, with a maximum temperature of 25 °C and a minimum of 9.2 °C. However, solar protection must be taken into account, in view of possible high exposures in summer and winter, when solar radiation is present. In Fig. 16.4b, the graph shows the variability of rainfall, remaining relatively constant during most of the year, the nights are cold with temperatures between 9° and 10 °C.

Winds and Psychometric Abacuses. The most frequent winds come from the south-west and the fastest from the south. The highest wind during the day has a speed of 5.6 m/s and during the night the speed is 2.7 m/s.

Figure 16.5a shows wind variability. Figure 16.5b shows that Moquegua is largely within the comfort zone, its climate is moderate with relative humidity and a large thermal amplitude during the year. The local population is in comfort from early morning and midday.

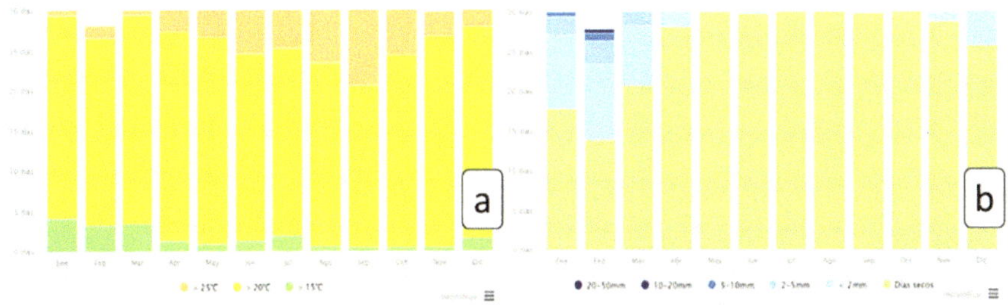

Fig. 16.4 Temperature and precipitation

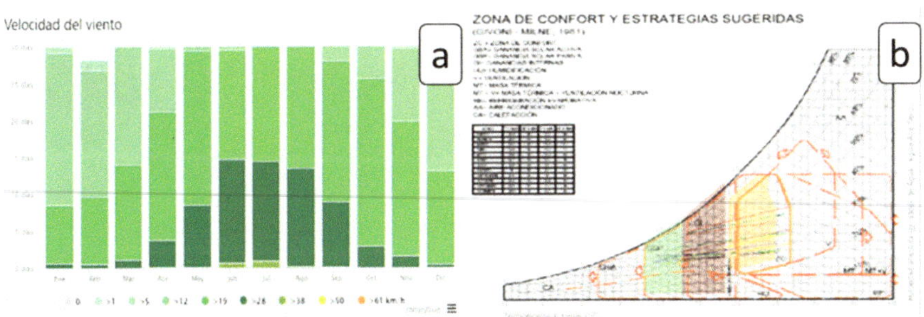

Fig. 16.5 Winds

16.2.4 Flora and Fauna

Flora. Moquegua has a total of 1226 taxa grouped in 3 groups, 105 families and 490 genera. These results give us the information and show the great quantity and diversity of flora that exists in the department of Moquegua. In the flora we find divisions such as the Magnoliophyta is the best represented with 96.5% (1183 spp.), another division we have the Pteridophyta with 3.0% (37 spp.) and finally the Pinophyta with 0.5% (6 spp.). Within Magnoliophyta, the Magnoliopsida class is the most abundant in the subject with 948 species, while on the other hand the Liliopsida presents 235 taxa (Siar.minam.gob.pe/moquegua/documentos/diagnostico-diversidad-biologica-region-moquegua).

Table 16.1 shows the division of species, genera and families of flora diversity.

Figure 16.6 shows the distribution of the flora of Moquegua. We can see in Table 16.2 the classification of species, genera and families that exist in the department of Moquegua.

Table 16.1 Flora classification

	Pteridophyta	Pinophyta	Magnoliophyta		Total
			Liliopsida	Magnoliopsida	
Species	37	6	235	948	1226
Genres	22	3	96	368	489
Families	10	3	15	77	105

Fig. 16.6 Distribution of flora

Table 16.2 Classification of fauna

Classes	Species	Genres	Families	Orders
Mammalia	38	28	15	7
Birds	163	106	43	19
Sauropsida	13	5	4	1
Aphibia	6	3	3	1
Total	220	142	65	28

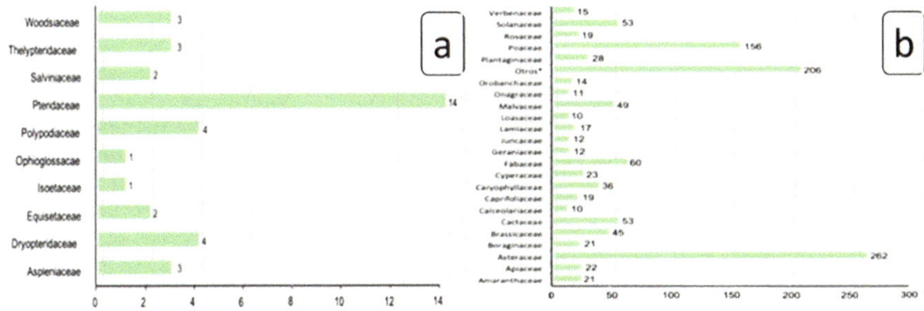

Fig. 16.7 Graph of species (**a**) and (**b**)

Figure 16.7a shows how the Pteridophyta species of Moquegua is divided, and Fig. 16.7b shows how the Magnoliophyta species of Moquegua (b) is divided

Fauna. Moquegua has a total of 220 species of different animals, these are grouped distinctively in 4 classes, 28 orders, 65 families and 142 genera. Here we realize that the best represented animal class is Birds with 74.1%, then mammals represent 17.3%, then reptiles 5.9% and finally amphibians 2.7%, Details in the following Table 16.2 (Siar.minam.gob.pe/moquegua/documentos/diagnostico-diversidad-biologica-region-moquegua).

Table 16.2 shows the species groups, genera, and families of faunal diversity.

Mammals and Birds. Moquegua has recorded a total of 38 species of mammals grouped into 15 families and seven taxonomic orders. Rodents have the largest number of species. These species belong to three families (Muridae, Cricetidae and Chinchillidae). Carnivores constitute the second group in species richness, with 12% of the total.

A total of 163 bird species were recorded in Moquegua, composed of 43 families and 19 taxonomic orders. The orders Passeriformes and Charadriiformes had the highest number of species, with each order recording 74 and 24 species, respectively.

Figure 16.8a shows the number of mammal species by order of origin. Figure 16.8b shows the number of bird species by order of origin

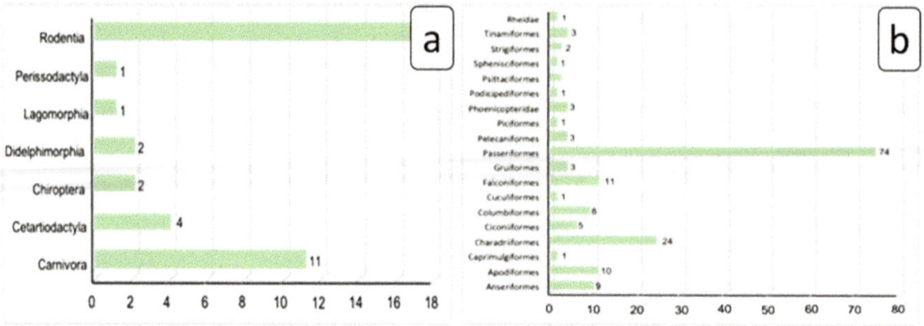

Fig. 16.8 Bird classification chart

Figure 16.9 shows the distribution of the fauna of Moquegua, we can see the classification of the fauna by species, genera and families that exist in the department of Moquegua.

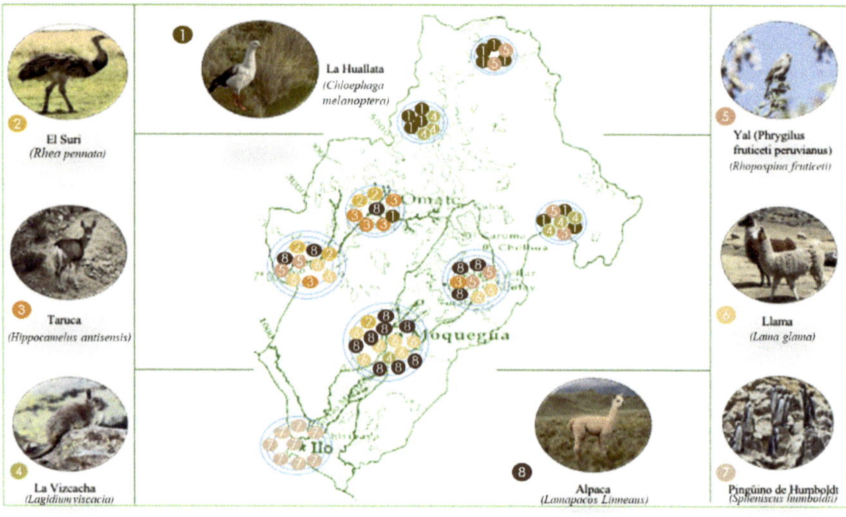

Fig. 16.9 Distribution of fauna

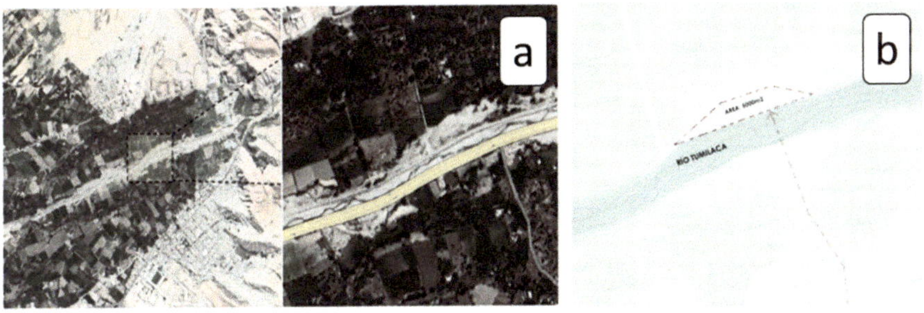

Fig. 16.10 Location map

16.3 Results

16.3.1 Project Location

The chosen lot is located on the banks of the Tumilaca River in the Moquegua region. It has an intervention area of 4300 m^2. It is currently considered an area in disuse and without vegetation, which allows to generate a public space to enhance the place. It has connecting roads with the city of Somague and Moquegua, which achieves a connection between the city and the rural area.

Figure 16.10a shows the location of the intervention area in a distant view and a close-up view. Figure 16.10b shows the accessibility of the project, which consists of a pedestrian bridge coming from the city of Samegua.

16.3.2 Volumetric Proposal

Figure 16.11a shows the zoning of the project, which consists of spaces such as the central plaza, sports areas, bicycle paths, trails, and a viewpoint within the public space on the Tumilaca River. Figure 16.11b shows the proposed route through ramps and trails, which is accessible to all.

Central Plaza

This square is proposed as an articulating axis of the proposal, and a space of union for the recreational activities of the sector.

Sports Areas

A space dedicated to sports was included to activate the user's productive character in the natural environment of Moquegua.

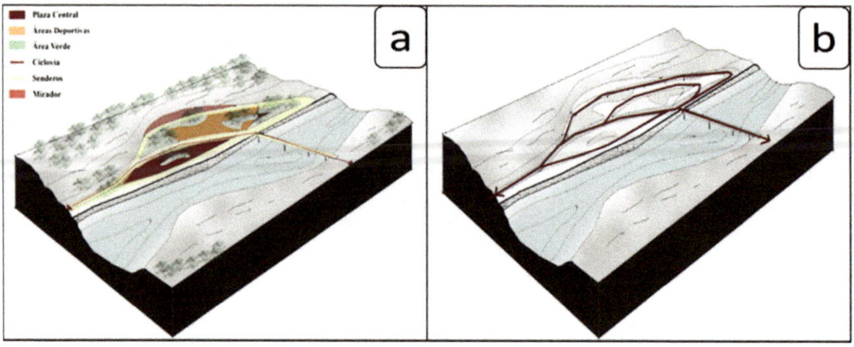

Fig. 16.11 Zoning and access

Green Area

Vegetation is used as an element of thermal comfort in each environment, as well as to generate free areas.

Ciclovía

Bicycle lanes are generated to connect with nearby cities, increasing accessibility to the project.

Trails

We propose tours through the natural environment offered by the site and a dynamic tour in the proposed public space.

Lookout

Within the proposal, a viewpoint was integrated where the user can visualize a better panorama and natural landscape that is obtained within the area.

16.3.3 Materiality

Figure 16.12 shows the materials implemented in the project, such as the trails and the permanent furniture. Wood will be a sustainable material, since emissions are 30% lower over a period of 50 years, saving money and benefiting the environment.

Figure 16.13 shows the system implemented on the riverbank, which consists of creating a gabion retaining wall to prevent overflows or flooding.

Figure 16.14 shows the different sub-levels of the project, which is coupled to the natural soil, as well as the solar path and wind orientation with respect to the project.

Figure 16.15a shows the gazebo, where the strategy of a wooden sun and shade cover is developed, which fulfills the function of greater thermal comfort and as a visual frame of the gazebo itself. Figure 16.15b also shows the furniture strategy in

Fig. 16.12 Use of textures in the project

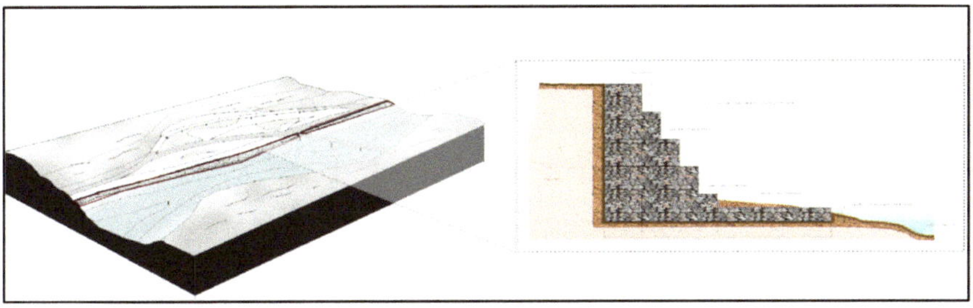

Fig. 16.13 Gabion system on the riverbank

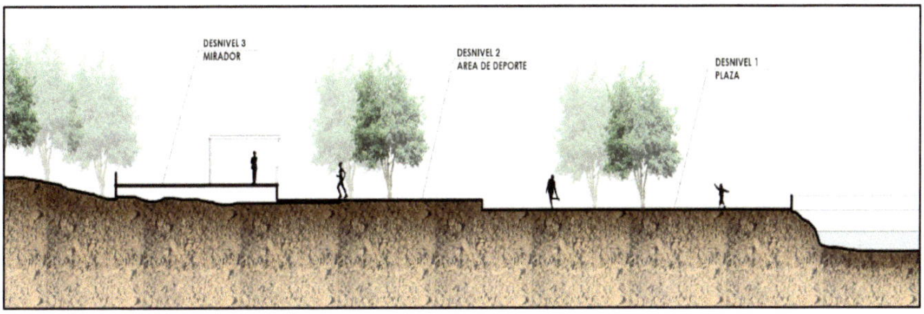

Fig. 16.14 Project slopes

the central plaza of the project, which uses the tree as a shade element, thus generating permanence in the space.

Figure 16.16 shows the use of sports furniture, accompanied by vegetation as an element of protection and comfort.

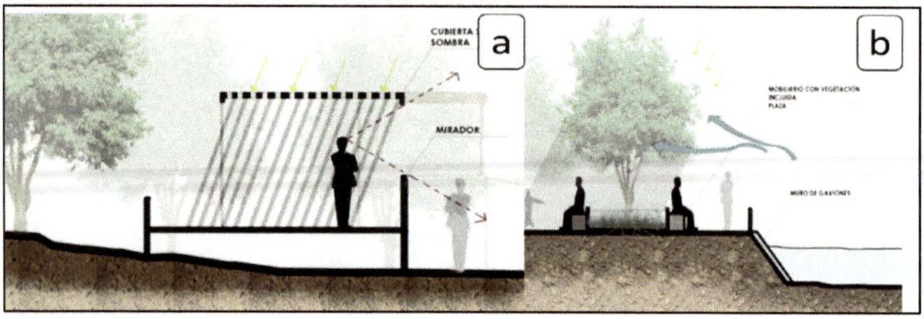

Fig. 16.15 Project slopes

Fig. 16.16 Productive user

16.3.4 Solar Path Analysis

Figure 16.17 shows the graph of the solar path with respect to the project at 12:00 pm.

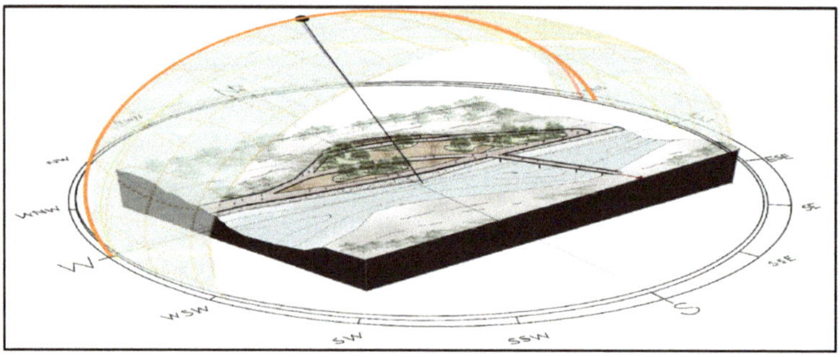

Fig. 16.17 Solar path

16.3.5 Proposed Planimetry

Figure 16.18 shows the general plan of the proposal, which consists of a plaza, sports area and viewpoint, together with bicycle paths and trails.

Figure 16.19 shows a cross section of the project, which shows the viewpoint, the sports area and the central plaza, as well as the pedestrian bridge.

Figure 16.20 shows a cross section of the project, which shows the viewpoint, the sports area and the central plaza, as well as the pedestrian bridge.

Figure 16.21a shows the entrance to the riverside public space, which consists of a pedestrian bridge connecting the project with the city of Somague. Figure 16.21b shows the square, where the furniture used, and the rest areas are shown.

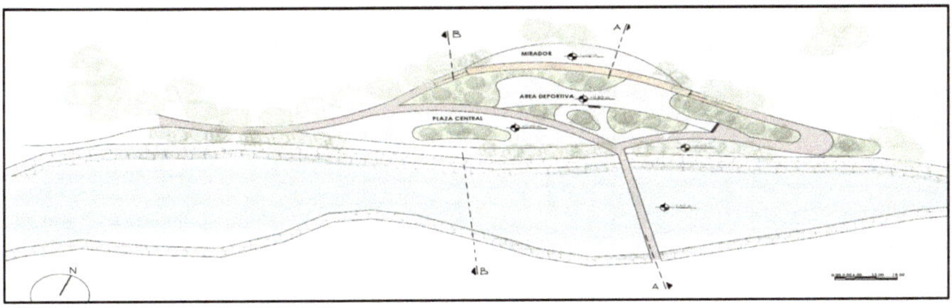

Fig. 16.18 Draft plan

Fig. 16.19 Cutting project

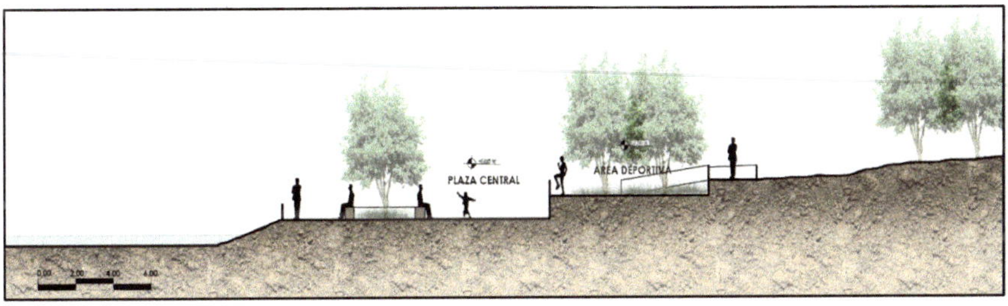

Fig. 16.20 Cutting project

Figure 16.21c shows the sports area, where the furniture used is placed near the trees to provide shade. Figure 16.21d shows the proposed roof for the upper gazebo, which uses the wooden pergola as a base and solar panel. Figure 16.21e shows the proposed furnishings for the small standing areas along the project, which are closer to the trees. Figure 16.21f shows the general view of the project, showing the gabion system implemented at the river's edge as a flood prevention strategy.

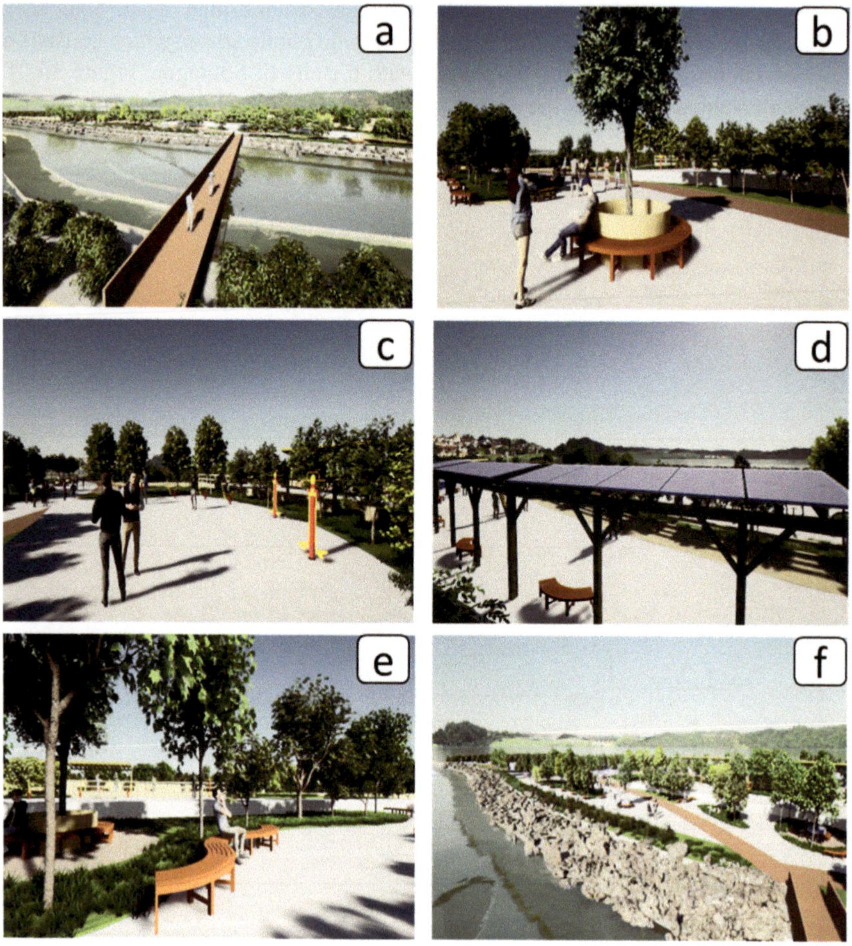

Fig. 16.21 Plaza

16.4 Discussion

Rivers are key to the location of cities, towns and urban growth. The exploration in the reports on risks and urbanism guides us to be able to raise proposals related to urban rivers, the contribution to a new urban perspective and a tracer of riparian treatments. The aforementioned background allows us to identify outstanding ideas of organization and design. Being the beginning to propose strategies that are linked to urban riparian designs, through their interventions (https://www.researchgate.net/publication/45647145_Efectos_ambientales_de_la_urban-izacion_de_las_cuencas_de_los_rios_Biobio_y_Andalien_sobre_los_riesgos_de_in-undacion_y_anegamiento_de_la_ciudad_de_Concepcion; Esenarro et al. 2022a). The intervention in urbanized watersheds, so that the results help decision makers to generate projects that mitigate or minimize the damage caused by floods. It was applied to the particular conditions of different coastal cities in the world, managing to represent vulnerability through bar graphs with normalized indices between 0 and 1. They constructed vulnerability indices applied to the coastline of the Basilicata region in southern Italy, generating vulnerability maps associated with four classes: very high, high, moderate and low. Both models are complementary (Análisis de riesgo por Inundación: Metodologia y aplicacion a la cuenca de Atemajac; Esenarro et al. 2022b).

16.5 Conclusion

The sustainable architecture proposal influences the design strategies that affect the study area, allowing to minimize the effects of climate change.

Likewise, the proposal for the regeneration of the Tumilaca river embankment improves the quality of life of the inhabitants, generating an interaction between them and their visual comfort.

The temperature of Moquegua provides us with scenarios to solve the problems of air conditioning, taking into account solar protection against possible high exposures in summer and winter, when solar radiation is present.

The proposed clean energies allow us to mitigate environmental pollution and take advantage of the local climate.

References

Adhikari D, Li J (2013) Modelling ambiguity in urban planning. Ann GIS 19:143–152

Alberto L, Aquino A (2016) El Espacio Público en la ribera del Río Piura para mejorar la calidad urbanística y paisajística de la zona: caso del malecón Miraflores, Piura

Análisis de riesgo por Inundación: Metodologia y aplicacion a la cuenca de Atemajac

Balsas C (2021) Sustainable urbanism: riverfront greenway planning from tradition to innovation. Soc Sci Res. https://doi.org/10.1080/13511610.2021.1920001
Cengiz B (2013) Urban river landscapes. In: Özyavuz M (ed) Advances in landscape architecture. InTech, London, UK, pp 551–586
Centro de Estudios y Prevención de Desastres (2003) Plan Director de Moquegua - Samegua 2003–2010. Zonificación de Riesgos. Moquegua: Municipalidad Provincial Mariscal Nieto
Danjaji AS, Ariffin M (2017) Green infrastructure policy for sustainable urban development. International article. Int J Environ Sustain Dev 16(2):112–127
Dickson E, Baker J, Hoornweg D, Tiwari A (2012) Urban risk assessments: an approach for understanding disaster and climate risk in cities. The World Bank, Washington, WA, USA
Esenarro SR, Raymundo V, Nizama PC, Huaman CR, Llanos WM (2022a) Use of solar energy in the design of the interpretive trail of the Antankallo waterfall in the district of Matucana, Peru 2022. In: International conference on electrical, computer and energy technologies (ICECET), Prague, Czech Republic, pp 1–6. https://doi.org/10.1109/ICECET55527.2022.9872565
Esenarro D, Malpartida K, Silvana L, Raymundo V, Morales W (2022b) Use of renewable energies applied in design strategies for user comfort in a house in Iquitos-Belen. In: 2022 11th international conference on power science and engineering (ICPSE), Eskisehir, Turkey, pp 135–141. https://doi.org/10.1109/ICPSE56329.2022.9935461
Fitzgerald J (ed) (2010) Emerald cities: Urban sustainability and economic development. Oxford University Press Inc., New York, NY, USA
Ganis M (2015) Planning urban places: self-organising places with people in mind. Routledge, New York, NY, USA
Hernández-Uribe RE, Barrios-Piña H, Ramírez A (2017) Análisis de riesgo por Inundación: Metodología y aplicación a la cuenca Atemajac
https://www.researchgate.net/publication/45647145_Efectos_ambientales_de_la_urbanizacion_de_las_cuencas_de_los_rios_Biobio_y_Andalien_sobre_los_riesgos_de_inundacion_y_anegamiento_de_la_ciudad_de_Concepcion
Kuon Cabello LE (2015) Retazos de la Historia de Moquegua
Levy JM (ed) (2016) Contemporary urban planning, 7th edn. Routledge, New York, NY, USA
Ministerio de Vivienda, Construcción y Saneamiento (2016) Plan de desarrollo Urbano y Sostenible Moquegua
Ministerio de Desarrollo Agrario y Riego pág. 14. https://www.midagri.gob.pe/
Pandit P, Fauggier G, Gu L, Knöll M (2020) How do people use Frankfurt Mainkai riverfront during a road closure experiment? A snapshot of public space usage during the coronavirus lockdown in May 2020. Cities Health 2020
Portner K (1992) Cooperative riverfront planning: the colorado river, grand junction Colorado. Can Water Resour J 17:264–269
Siar.minam.gob.pe/moquegua/documentos/diagnostico-diversidad-biologica-region-moquegua
Steiner FS (2018) Making plans: how to engage with landscape, design, and the urban environment. University of Texas Press, Austin, TX, USA

Part IV
Traffic Management and Related Carbon Emission Analysis

Chapter 17
Calculation of Carbon Emissions from Urban Passenger Transport Based on Decision Tree

Jinli Wei, Liuying Lu, Linhao Zhang, and Mengmeng Zhang

Abstract In order to scientifically and accurately quantify the carbon emissions of urban passenger transportation, a carbon emission calculation model based on residents' travel decisions and the IPCC "bottom-up" calculation method was proposed. Make full use of multi-source data, select the actual CO, HC, NOx and CO_2 emission factors, establish a decision tree model, simulate the process of residents' selection of different travel modes and corresponding fuels, use multi-logit models to calibrate the selection probability, and take Jinan City as an example to verify the reliability of the model. The research conclusion have shown that the model comprehensively considers the influence of motor vehicle carbon emission parameters and the randomness of residents' travel choices, and the fitting accuracy is 92.37%,which can effectively measure the carbon emission of urban passenger transportation. This method has wide applicability and can provide theoretical and empirical reference for the measurement of carbon emissions from passenger transportation in other cities and the realization of low-carbon transportation modes.

Keywords Environmental engineering · Decision tree · Passenger transport · Carbon emission · Multi-logit Model

17.1 Introduction

With the rapid advancement of urbanization, the number of motor vehicles in my country continues to increase, and the transportation sector has become the fastest growing sector of GHG emissions (International Energy Agency 2016). Currently, Many scholars have done a lot of research on carbon emission measurement. Kumar

J. Wei · L. Lu (✉) · L. Zhang
Qingdao University of Technology, Qingdao 266520, China
e-mail: 1483458311@qq.com

M. Zhang
Shandong Jiaotong College, Jinan 250300, China

C. Yuan et al. (eds.), *Proceedings of 4th International Conference on Resources and Environmental Research—ICRER 2022*, Environmental Science and Engineering,
https://doi.org/10.1007/978-3-031-31808-5_17

MK (Kumar 2016) used the "bottom-up" method to obtain a total of 2.12 million tons of CO_2 in Chennai, of which the transportation sector accounted for 29.7%; Kakouei A (Kakouei and Idris 2012) concluded that the CO_2 emissions from private cars in Tehran accounted for 88%. Jia Peipei (Peipei 2018) calculated Xi'an City according to the "bottom-up" method, and found that more than 60% of CO_2 comes from private cars. The application of decision tree method in transportation has also become widespread. Niu Kai (Kai 2021) constructed a decision tree model for the reachability process and the car ride process; Zhu Yao (Zhu Yao 2021) used the decision tree model in the process of license plate recognition Ying Zimin (Zimin et al. 2018) proved that residents' travel decisions are the fundamental factors that affect the carbon emissions of residents' travel and transportation, and obvious differences in the carbon emissions.

However, in the actual calculation process, the unified default value provided by IPCC is often directly used, resulting in large differences in calculation results, which can't truly reflect the emission status of different cities. At the same time, few studies have applied the decision tree method to carbon emissions measurement, ignoring the impact of residents' decision-making on carbon emissions, therefore, a carbon emission calculation model based on residents' travel decisions and the IPCC "bottom-up" calculation method was proposed.

17.2 Modeling of Carbon Emission Estimation Based on Decision Tree

17.2.1 Modeling Idea

Carbon emission of urban passenger transportation is related to the level of motor vehicles and a significant relationship with the travel choices of residents. Therefore, this study will construct a carbon emission model of urban passenger transportation structure from the above two perspectives. Using the decision tree and to combine the IPCC "bottom-up" calculation method, establish a carbon emission measurement model to directly reflect the carbon emissions of different travel modes.

Decision trees are generally generated from top to bottom, and select the decision attribute with the best classification effect at each node, until can accurately classify the sample. Then intuitively display the generation process and results. According to the method characteristics of the decision tree, to reflect the travel choices of residents, the types of urban passenger transportation are mainly divided into private car, bus, rail, and taxi. For each travel mode, classify the fuel as the probability branch in the decision tree, and P_i (resident travel probability) is calibrated by a multi-logit model. Considering the motorcycles difficulty of obtaining data, it is not considered. Figure 17.1 shows the construction of a decision tree model for a resident's one-time trip.

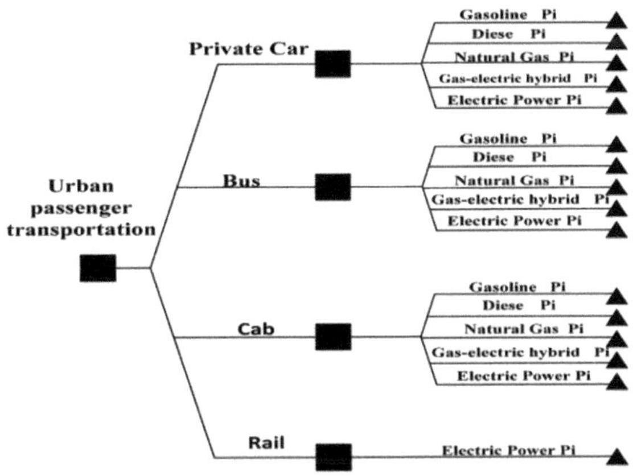

Fig. 17.1 Resident travel decision tree model

17.2.2 Model Establishment

The main pollutants emitted by motor vehicles include the following six types: CO, CO_2, HC, NO_x, O_3, PM, VOC. Among them, CO、CO_2、NO_X、HC these four pollutants have the most serious impact on the atmospheric environment. Therefore, the "carbon" in this paper refers to common greenhouse gases, namely CO, HC, NO_X and CO_2.

The carbon emission accounting method can be known from the "bottom-up" calculation method (Xiaoli 2017), as shown in formula (17.1).

$$E_{pt} = \sum\nolimits_{i,j,k} (N_{i,j,k} \times D_{i,j,k} \times e_{i,j,k}) \tag{17.1}$$

In the formula, "E_{pt}" is the total carbon emission of urban passenger transportation, Kg; "N" is the vehicle ownership, vehicle; "D" is the average annual mileage, km; "e" is the emission factor, kg/km; "i" is the vehicle type; "j" is the fuel type; "k" represents the pollutant type.

Considering the randomness of residents' choice, a carbon emission model is constructed, shown in formula (17.2).

$$E_{pt} = \sum\nolimits_{i,j,k} (N_{i,j,k} \times P_{i,j} \times D_{i,j,k} \times e_{i,j,k}) \tag{17.2}$$

In the formula, "P" is the probability that residents choose vehicle type i to use fuel j.

Among them, the carbon emission calculation formula of electric vehicles and rail transit is shown in formula (17.3).

$$\mathrm{E_{ele,R} = N_{i,k} \times P_i \times D_{i,k} \times C_{i,k} \times e_{ele,i,k}} \tag{17.3}$$

In the formula, "$E_{ele,R}$" is the total carbon emission of electric vehicles and rail transit, Kg; "C" is the power consumption per 100 km, kWh/100 km; "$e_{ele,i}$" is the average CO_2 emission factor, kgCO_2/kWh; "i" is the vehicle type; "k" represents the pollutant type.

17.3 Parameter Calibration

Paper focuses on the calibration of two parameters, the emission factor and the residents' travel choice probability.

17.3.1 Emission Factor

As an existing vehicle emission data source, vehicle annual inspection data has the high reliability (Shunying and Qiucheng 2020). The calibration of the emission factor follows the principle that the annual inspection data is superior to the literature data.Extract the sample size of the annual inspection report that meet the 95% level of the total annual inspection data sample to obtain the emission factors of CO, HC and NO_x, g/km; the detected process data obtains the instantaneous emission quality of CO_2, g/s, which can be converted into g/km by formula (17.4).

$$e = \frac{3600R}{V} \tag{17.4}$$

In the formula, "e" is the emission factor, g/km; "R" is the instantaneous emission mass, g/s; "V" is the driving speed, km/h.

17.3.2 Residents' Travel Choice Probability

The choice of fuel choice behavior for each travel mode has the characteristics of discreteness and no order, and the number of options is greater than 3, so the multi-logit model (WANG Jia., LIU Jingbo., LIU Yongda. 2020) can be used to calibrate the choice.

Multinomial logit model The multi-logit model, the decision-maker will choose the option with the greatest utility. Its theoretical basis is stochastic utility theory. As is shown in Formula (17.5).

$$U_{m,n} = V_{m,n} + \varepsilon_{m,n} \tag{17.5}$$

In the formula, "$U_{m,n}$" is the utility function of the choice of branch "m" for traveler "n", and "$V_{m,n}$" is the observable fixed part of the choice of branch "m" for traveler "n"; "$\varepsilon_{m,n}$" is Traveler "n" to choose the unobservable random perturbation part of branch "m". "$V_{m,n}$" is shown in formula (17.6).

$$V_m, n = \alpha^i + \beta_1^i x_1 + \beta_2^i x_2 + \ldots\ldots + \beta_z^i x_z \tag{17.6}$$

In the formula, "α^i" is the characteristic variable; "α^i" and "β_z^i" is the weight parameters of each influencing factor, which are obtained by maximum likelihood estimation.

It is assumed that "$\varepsilon_{m,n}$" obeys a double exponential distribution with parameters (λ, ω) that are independent of each other. Usually take $\lambda = 0$. The distribution function is shown in formula (17.7).

$$F_\varepsilon(x) = \exp[-\exp(-\omega(x - \lambda))], \omega > 0 \tag{17.7}$$

The probability of individual chooses to choose branch m is shown in formula (17.8).

$$\mathrm{P_{m,n}} = \frac{\exp(\alpha^i + \sum_{z=1}^{n} \beta_z^i x_z)}{\sum_{m=1}^{n} \exp(\alpha^i + \sum_{z=1}^{n} \beta_z^i x_z)} \tag{17.8}$$

In the formula, $0 \le P_{m,n} \le 1, = \sum_{z=1}^{n} p_{m,n}$ 1。

Model Construction The questionnaire results were obtained by face-to-face random SP travel survey of residents, which ensured the validity of the questionnaire. In terms of sample size, referring to Peduzzi's research, the minimum sample size is at least 10 × The ratio of the number of independent variables to the number of minimum results (Peduzzi and Kemper 1996). The fuel selection of each travel mode in the questionnaire is coded as 1 (gasoline), 2 (diesel), 3 (natural gas), 4 (gas electric hybrid), 5 (electricity). The basic characteristics of each sample are independent variables of logit model, and are classified and allocated (Longyu 2018), as shown in Table 17.1.

The dependent variable of the Logit model is the fuel choice of a certain travel mode, represented by P_1, P_2, P_3, P_4, P_5, and the reference category is the choice of gasoline, so the utility function of the travel decision maker's choice is as shown in Eq. (17.9)-Eq. (17.12) as shown:

$$V_2 = In\frac{P_2}{P_1} = \alpha^2 + \beta_1^2 x_1 + \beta_2^2 x_2 + \ldots\ldots \beta_8^2 x_8 \tag{17.9}$$

$$V_3 = In\frac{P_3}{P_1} = \alpha^3 + \beta_1^3 x_1 + \beta_2^3 x_2 + \ldots\ldots \beta_8^3 x_8 \tag{17.10}$$

$$V_4 = In\frac{P_4}{P_1} = \alpha^4 + \beta_1^4 x_1 + \beta_2^4 x_2 + \ldots\ldots \beta_8^4 x_8 \tag{17.11}$$

Table 17.1 Classification table of characteristic variables

Classification	Characteristic variable	Assign
Personal attributes	Gender X_1	1 (Male) 2 (Female)
	Age X_2	1 (7 ~ 20) 2 (21 ~ 45) 3 (45 ~ 60) 4 (>60)
	Profession X_3	1 (Student) 2 (Civil servant) 3 (Business) 4 (other)
	Low carbon travel understanding X_4	1 (Totally no idea) 2 (Heard about it) 3 (Very clear)
	State subsidy X_5	1 (Yes) 2 (NO)
Family attributes	family population X_6	1 (≤3) 2 (>3)
	monthly income X_7	1 (3 ~ 6) 2 (6 ~ 10) 3 (>10)
Travel attributes	Economy X_8	1 (Very dissatisfied) 2 (Dissatisfied) 3 (common) 4 (Satisfaction) 5 (Very satisfied)

$$V_5 = In\frac{P_5}{P_1} = \alpha^5 + \beta_1^5 x_1 + \beta_2^5 x_2 + \ldots\ldots \beta_8^5 x_8 \tag{17.12}$$

17.4 Instance Verification

17.4.1 Data Sources

This paper takes Jinan as an example to verify. The number of motor vehicles comes from the Yearbook (China Statistics Press.: Jinan Bureau of Statistics 2020); The average annual driving mileage of motor vehicles is the average value of the Technical Manual for the Preparation of the Emission List of Urban Air Pollution Sources[14]and the literature (Yimeng 2020; Dong et al. 2021); Calculate the emission factor according to 3.1 to determine. A total of 1500 questionnaires were distributed to the residents of Jinan City, and 1339 valid questionnaires were finally recovered, with an effective rate of 89.26%, meeting the sample size requirements. The sample size can effectively support the research results.

17.4.2 Emission Factor "E" Calibration

The emission factors from the annual inspection data and literature data (Jiguang 2021; Pan Yingjiu (2020) as shown in Table 17.2.

17.4.3 Parameter Calibration of Logit Model

Fuel choices that are not significant for the mode of transportation are considered as null variables. The results show that the model likelihood ratio test has a significant level of Sig. = 0.014 < 0.05, It means that the variables have a significant effect on the composition of the model and the model is valid. The characteristic variables of whether or not to know about low-carbon travel (X_4), the availability of state subsidies (X_5), personal monthly income (X_7), and travel economics (X_8) have a significant effect on the fuel choice of transportation mode. When calibrating, only the travel modes and factors with significant influence are listed. The residents' choice of fuel for private car travel is explained in detail. The rest of the modes are similar. Therefore, the utility parameters of diesel, natural gas, gas-electric hybrid, and electric power are calibrated by gasoline private cars as the reference category. The model parameters are calibrated as shown in Table 17.3.

Substitute the parameter values of each category into Eqs. (17.9–17.12) to obtain the selection probability of each travel mode, as shown in Table 17.4.

17.4.4 Analysis of Results

Substitute various parameters into Eq. (17.2) to obtain the total carbon emissions of urban passenger transportation in Jinan City, as shown in Fig. 17.2.

According to the total carbon emissions, the carbon emission structure and the carbon emission structure of different fuels can be obtained, as shown in Figs. 17.3, 4.

As can be seen from Fig. 17.2, the total emissions showed a trend of increasing, but an inflection point occurred in 2018, mainly due to the rapid increase in electric buses.

From Fig. 17.3, the contribution rate of carbon emissions in passenger transportation in Jinan City is: private car > taxi > bus > rail.

From Fig. 17.4, the main source of carbon emissions from urban transportation is the use of traditional fuels, accounting for more than 90%.

The carbon emission structure of different fuels for different transportation modes was calculated. The results are shown in Figs. 17.5, 17.6 and 17.7.

Table 17.2 Emission factors of various pollutants

Mode of transportation	Fuel type	Emission factor /g·km^{-1}			
		CO	NO_x	HC	CO_2
Private car	Gasoline	1.09	0.22	0.13	195.48
	Electricity	–	–	–	70 (Jiguang 2021)
Cab	Gasoline	1.19	0.17	0.16	176
	Natural Gas	1.03	0.15	0.10	200
	Gas-electric hybrid	0.62	0.07	0.04	164.18
Bus	Diesel	3.65 Jiguang (2021); Pan Yingjiu (2020),	15.1 Jiguang (2021); Pan Yingjiu (2020),	1.385 Jiguang (2021); Pan Yingjiu (2020),	910
	Electricity	–	–	–	703.5 g/kwh Jiguang (2021)
	Natural Gas	18 Pan Yingjiu (2020)	0.90 Pan Yingjiu (2020)	9.3 Pan Yingjiu (2020)	3 600 Pan Yingjiu (2020)
Rail	Electricity	–	–	–	860 g/kwh Pan Yingjiu (2020)

Table 17.3 Parameter calibration results of SPSS software

Diesel			
Independent variable	Parameter estimation	Wald	Significant level
intercept	2.168	3.405	0.000
$X_4 = 1$	1.792	3.283	0.024
$X_4 = 2$	6.797	10.719	0.001
$X_4 = 3$	9.109	7.076	0.003
$X_5 = 1$	4.676	21.197	0.008
$X_5 = 2$	0.733	3.709	0.023
$X_7 = 1$	−5.386	7.451	0.000
$X_7 = 2$	1.226	6.212	0.012
$X_7 = 3$	−1.044	10.129	0.000
$X_7 = 4$	0.204	12.711	0.000
$X_8 = 1$	−6.484	11.726	0.009
$X_8 = 2$	3.108	16.543	0.013
$X_8 = 3$	2.108	17.753	0.005
$X_8 = 4$	0.921	15.682	0.021
$X_8 = 5$	0.414	13.043	0.000
Gas-electric hybrid			
Independent variable	Parameter estimation	Wald	Significant level
intercept	−1.667	2.852	0.021
$X_4 = 1$	2.207	5.197	0.043
$X_4 = 2$	7.454	15.216	0.010
$X_4 = 3$	10.174	14.236	0.001
$X_5 = 1$	3.559	11.134	0.008
$X_5 = 2$	6.491	13.725	0.011
$X_7 = 1$	−0.835	19.097	0.035
$X_7 = 2$	1.236	6.667	0.000
$X_7 = 3$	4.647	13.269	0.015
$X_7 = 4$	1.206	5.246	0.014
$X_8 = 1$	4.053	13.226	0.022
$X_8 = 2$	1.117	4.526	0.000
$X_8 = 3$	7.895	3.214	0.020
$X_8 = 4$	−3.598	4.251	0.012
$X_8 = 5$	4.366	2.251	0.018
Natural Gas			
Independent variable	Parameter estimation	Wald	Significant level

(continued)

Table 17.3 (continued)

Diesel			
Independent variable	Parameter estimation	Wald	Significant level
intercept	−1.226	3.869	0.010
$X_4 = 1$	2.608	7.076	0.025
$X_4 = 2$	4.312	13.321	0.04
$X_4 = 3$	6.069	8.653	0.002
$X_5 = 1$	4.356	15.869	0.016
$X_5 = 2$	3.521	12.008	0.007
$X_7 = 1$	−4.864	10.719	0.003
$X_7 = 2$	3.433	13.256	0.001
$X_7 = 3$	5.291	12.268	0.000
$X_7 = 4$	−1.636	8.852	0.000
$X_8 = 1$	3.995	14.126	0.031
$X_8 = 2$	−1.482	11.205	0.016
$X_8 = 3$	5.366	21.202	0.023
$X_8 = 4$	−3.354	6.259	0.001
$X_8 = 5$	1.969	12.274	0.000
Electricity			
Independent variable	Parameter estimation	Wald	Significant level
intercept	2.572	14.186	0.003
$X_4 = 1$	2.608	7.076	0.025
$X_4 = 2$	4.788	3.215	0.019
$X_4 = 3$	9.265	7.158	0.006
$X_5 = 1$	4.676	15.869	0.016
$X_5 = 2$	1.622	4.987	0.000
$X_7 = 1$	4.864	10.719	0.003
$X_7 = 2$	−2.108	6.826	0.013
$X_7 = 3$	4.296	1.134	0.000
$X_7 = 4$	10.235	12.229	0.015
$X_8 = 1$	3.995	14.126	0.031
$X_8 = 2$	−7.714	10.256	0.006
$X_8 = 3$	4.416	6.227	0.000
$X_8 = 4$	4.131	1.960	0.001
$X_8 = 5$	8.216	6.229	0.000

Table 17.4 Probability of residents' travel choice

	Gasoline	Diesel	
Private car	0.970	0.003	
Bus	0.0024	0.0012	
Cab	0.327	0.001	
Rail	0.006	0.012	
	Natural Gas	Gas-electric hybrid	Electricity
Private car	0.004	0.010	0.013
Bus	0.963	0.0014	0.032
Cab	0.523	0.002	0.127
Rail	0.001	0.001	0.980

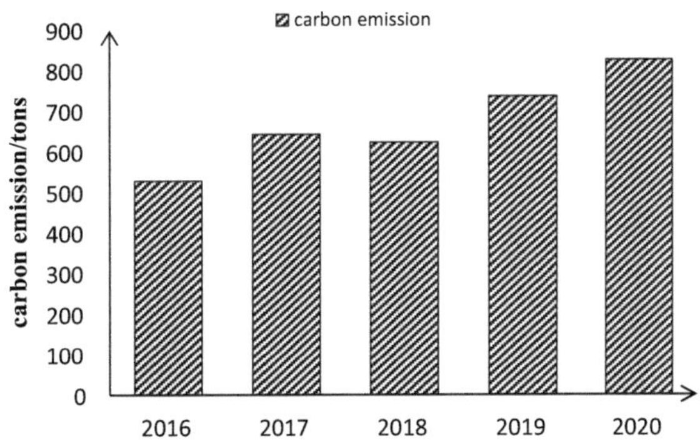

Fig. 17.2 Total carbon emission of urban passenger transport

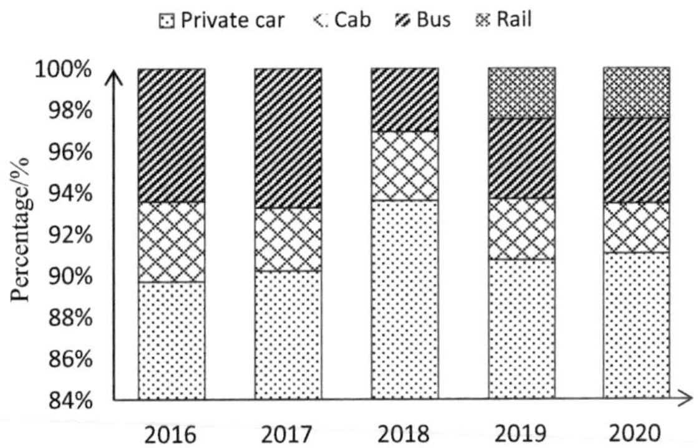

Fig. 17.3 2016–2020 Urban passenger transport emission structure

From Fig. 17.5, the total emissions of buses reached the lowest value in 2018, mainly because the number of diesel buses and natural gas buses decreased significantly.

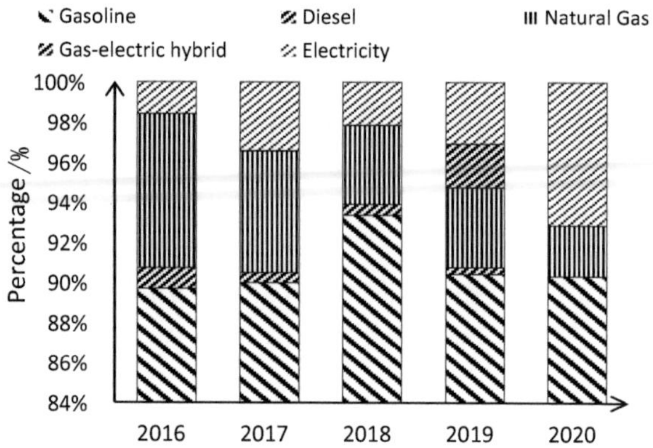

Fig. 17.4 carbon emission structure of different fuels

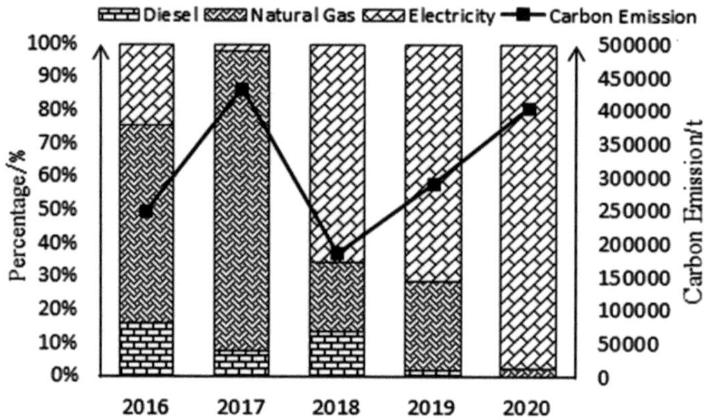

Fig. 17.5 Carbon emission structure trend of different fuels of buses

Fig. 17.6 Taxi fuel carbon emission structure trend chart

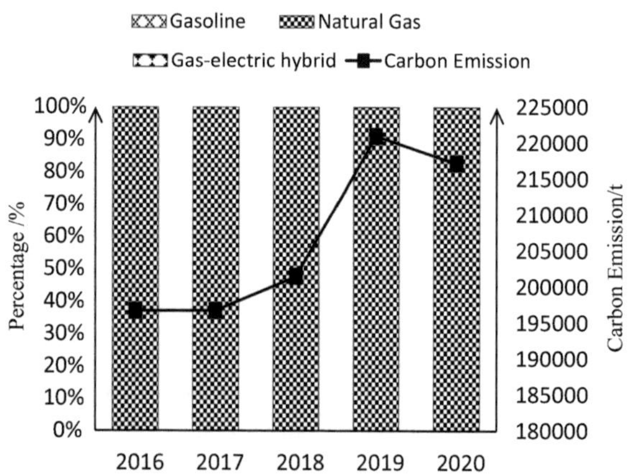

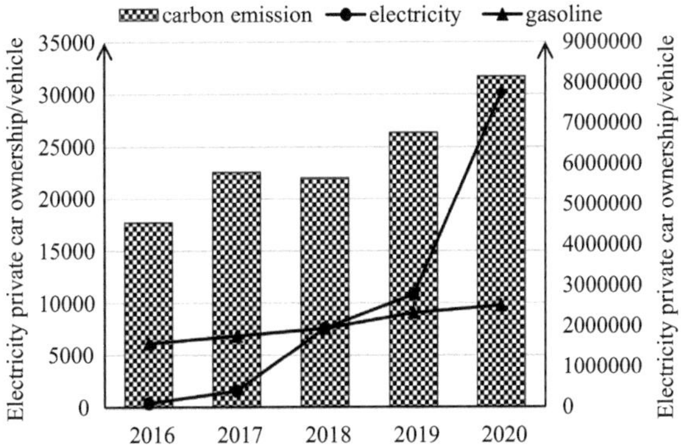

Fig. 17.7 Trend of carbon emission structure of different fuels for private car

From Fig. 17.6, the carbon emissions of taxis are also rising, except 2020, which is related to the new crown epidemic. Natural gas accounts for more than 99%, which is caused by using natural gas of taxis more than 95%.

From Fig. 17.7, it can be seen that the carbon emissions of private cars and ownership have a positive correlation trend, which is similar to the trend of fuel types.

17.4.5 Model Checking

Calculation results in this paper are compared with the calculation results of urban passenger transport carbon emissions in Jinan City from 2016 to 2020 in CEADs (Database 2020), the results are shown in Fig. 17.8. The results show that the error between the calculation results in this paper and the CEADs results is less than 10%, $R^2 = 0.9237$,and the overall simulation accuracy is good.

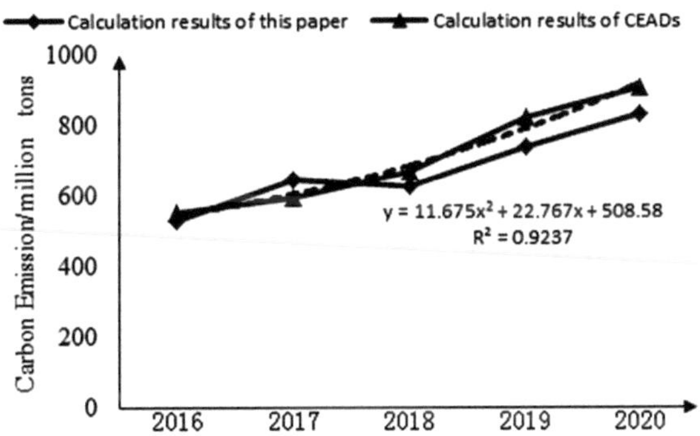

Fig. 17.8 Fitting diagram of the results of this paper and literature (Database 2020)

17.5 Conclusion

A carbon emission calculation model based on decision tree model and IPCC "bottom-up" calculation method is constructed. Taking Jinan City as an example, the main conclusions are as follows.

(1) The simulation accuracy of this model is $R^2 = 0.9237$, and the overall simulation accuracy is good.
(2) The calculation results of this model have high accuracy and authenticity. The emission coefficient shall be calibrated according to the annual inspection data of motor vehicles; The polynomial logit model is used to calibrate the trip probability of residents.
(3) This model comprehensively considers the impact of vehicle emissions and residents' choice on passenger transport carbon emissions measurement and can provide theoretical basis and practical reference for other cities.

References

China Statistics Press.: Jinan bureau of statistics. Jinan statistical yearbook. 2020 edn. China Statistics Press. Jinan (2020)

China Carbon Accounting Database, https://www.ceads.net.cn/, last accessed 2020.11.16

Fang Xiaoli., Luo Yong (2017) A comparative study on calculation methods of urban traffic carbon emissions. Transport Energy Conservation & Environmental Protection 13(4), 81–83

International Energy Agency.: CO_2 emission from fuel combustion 2016. IEA Publication, Paris (2016)

Kakouei A, Idris AKB (2012) An estimation of traffic related CO_2, emissions from motor vehicles in the capital city of. Iran. Journal of Environmental Health Science and Engineering 9(1):13

Kumar MK, Nagendra SMS.: Quantification of anthropogenic CO_2, emissions in a tropical urban environment. Atmospheric Environment 125, 272–282 (2016).

Ma Dong, Gao Jun, Liu Min (2021) Study on the characteristics of light duty vehicle kilometers traveled in typical cities based onbig date analysis. Transport Energy Conservation & Environmental Protection 17(6), 20–26

Ministry of Environmental Protection.: Technical manual for compilation of Emission inventory of urban air sources. Tsinghua university. Beijing (2015)

Niu Kai, Luo Ruiqi (2021) Residents' travel mode analysis based on the whole travel process. Traffic & Transportation 37(5), 18–23

Pan Yingjiu (2020) Analysis and estimation of emission characteristics of new energy buses based on multisource data. Southeast University. Nanjing

Peduzzi P, Kemper EA (1996) Simuation study of the number of events pervariable in logistic regression analysis. Journal of clinical epidemiology 49(12), 1373–1379

Peipei J (2018) Analysis and empirical study of the urban transport carbon emissions. Value Engineering 31(32):290–291

Shi Longyu, Sun Jing (2018) Study on the methods of assessment for low-carbon development of Chinesecities. Acta Ecologica Sinica 38(15), 5461–5472

Wang Jiguang, Gui Huaqiao (2021) Vehicle emission inventory and economic benefits of emission reduction in Tangshan. China Environmental Science 41(11), 5114–5124

WANG Jia., LIU Jingbo., LIU Yongda (2020) Research on the development model of urbanand rural road passenger transport integrationbased on rank analysis and multinomial Logit model. Journal of Railway Science and Engineering 17(12), 3224–3230

Wang Yimeng (2020) Study on greenhouse gas accounting and emission reduction potential of urban transportation. Hebei University of Engineering. Handan

Ying Zimin., Wu Xu., Yang Wu. (2018) Carbon emission accounting for the transition of public buses from gasoline to electricity in Hangzhou city. Acta Ecologica Sinica 38(18), 6452–6464

Zhu Yao (2021) Design of license plate number recognition based on decision tree. Automobile Applied Technology 46(22), 222–225

Zhu Shunying, Chen Qiucheng, Xiao Wenbin (2020) Marginal effect of Wuhan residents' travel mode under low carbon subsidy scenario. Journal of Safety and Environment 20(3), 1118–1125

Chapter 18
Joint Optimization of Production and Shipping for Small Orders Considering Less-Than-Container Loading

Yajing Tian, Guoqiang Wang, Yingying Chen, and He Luo

Abstract This paper introduces a joint optimization problem of production and shipping for overseas small orders considering less-than-container loading (LCL) in manufacturing enterprises. The problem consists of finding a production schedule together with a LCL shipping scheme simultaneously to minimize the total cost. An integer programming model is developed, and the genetic algorithm of enhanced elite reservation (SEGA) is used to solve large-sized instances. Besides, the performance of SEGA is compared with the classical genetic algorithm (SGA) and stallion genetic algorithm (stud GA). It is statistically shown that SEGA outperforms other algorithms in finding near-optimal solutions.

Keywords Small orders · Production · Less-than-container Loading (LCL) · Joint optimization

18.1 Introduction

With the growing trend of globalization, how to coordinate the task planning of orders under a decentralized production network is a normal problem for enterprises (Lohmer and Lasch 2021). Orders are divided into domestic and overseas orders, and the former is easier to handle, while the latter involves shipping and attracts less attention (Sun et al. 2018). When some overseas customers have few product requirements in their orders. A container needs to be filled with multiple small orders before it can be shipped. Although the product demand for each small overseas order is small, the number of orders is large, and the total manufacturing demand is huge

Y. Tian · G. Wang · Y. Chen · H. Luo (✉)
School of Management, Hefei University of Technology, Hefei, China
e-mail: luohe@hfut.edu.cn

Key Laboratory of Process Optimization & Intelligent Decision-Making, Ministry of Education, Hefei, China

C. Yuan et al. (eds.), *Proceedings of 4th International Conference on Resources and Environmental Research—ICRER 2022*, Environmental Science and Engineering,
https://doi.org/10.1007/978-3-031-31808-5_18

(Zhang et al. 2019). Therefore, it is crucial for manufacturing enterprises to determine the production plan and the LCL shipment scheme for small orders at the same time.

When it comes to the production and transportation of multiple factories, the assignment and scheduling problem of multi-factory operations involving inland and maritime transportation was studied by X.T. (Sun et al. 2015); (Feng et al. 2018). Proposed two heuristic algorithms for coordinating production and transportation planning (Pang 2019). Introduced a coordinated scheduling problem of distributed multi-plant production with the goal of minimizing the maximum delivery time. The problem of multi-plant production and distribution scheduling was studied by (Marandi and Ghomi 2019) to minimize delay and transportation costs (Sun and Lee 2015). Considered collaborative scheduling of production and distribution (Li and Li 2021). Studied the joint optimization of production and transportation under multiple factories, multiple production lines, and a single product considering the delivery date.

When it comes to task planning for small orders, (Kuhn et al. 2021) studied the order batching, picking, and distribution operations of small orders as a comprehensive planning problem (Lam et al. 2014). Combined mathematical model and fuzzy logic technology to divide small orders into batches and pick goods separately (Zhang and Ding 2018). Put forward an order group model for multi-variety and small-batch orders (Qin et al. 2022). constructed a joint optimization model of order allocation and path planning for small orders (Wang et al. 2017). constructed a dual-objective optimization model, which improved the processing efficiency of small orders.

However, the coordination of multi-factory production and transportation of small orders is particularly critical for manufacturing enterprises, especially when it comes to less-than-container loading (LCL) shipping. Therefore, this paper developed a joint optimization model of production and shipping for overseas small orders considering LCL. A genetic algorithm of enhanced elite reservation (SEGA) is developed, which provides a reference for manufacturing enterprises to deal with small orders.

18.2 Problem Description

There are n given small orders from the overseas customers and m factories are domestically located at different locations. The small orders will be shipped through K LCL schemes from R terminals to complete the delivery target of orders. The destination port of n small orders here is the same. Each LCL scheme includes information such as delivery time, shipment time, estimated time of arrival, available volume, unit shipping cost, terminal, and so on (see Fig. 18.1).

Each small order will not be split. Small orders assigned to the same factory all start production at the same time. The products will be immediately stored in the warehouse nearby the factory after they are finished. The delivery time and cost between the factory and warehouse are negligible (see Fig. 18.2).

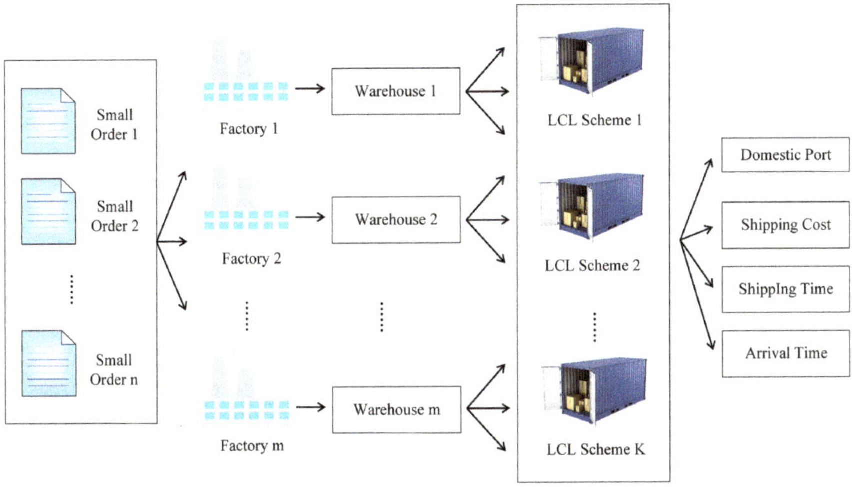

Fig. 18.1 The joint optimization scenario of production and shipping for overseas small orders considering LCL

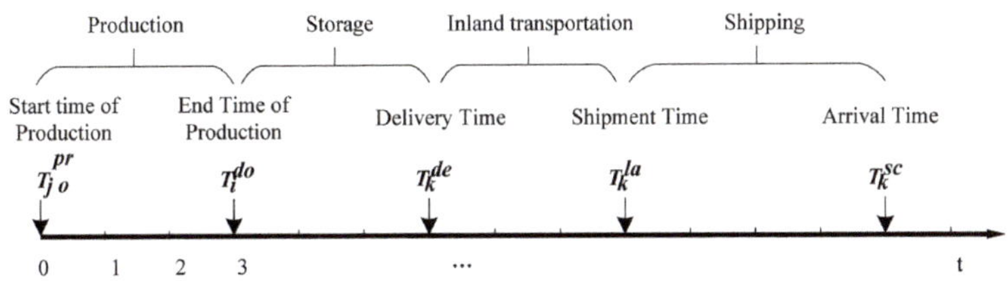

Fig. 18.2 The processing timeline for small orders

18.3 Mathematical Model

18.3.1 Nomenclature

The parameters and decision variables is shown (see Tables 18.1 and 18.2).

18.3.2 Objective Function

The objective function is to minimize the comprehensive cost of all small orders, which consists of production cost, storage cost, transportation cost, and penalty cost.

$$\min C = C^{pro} + C^{sto} + C^{tra} + C^{P} \tag{18.1}$$

Table 18.1 Parameters design and description

Parameters	Description
i	Index for small order, i=1,2,…, n
j	Index for factory, j=1,2,…, m
r	Index for terminal, r=1,2,…, R
k	Index for LCL scheme, k=1,2,…, K
V_i^{re}	Product volume of small order i
T_i^{re}	Deadline for small order i
Q_i	Product quantity of small order i
T_i^{Com}	Completion time of small order i
T_{max}	Maximum completion time for small orders
N_j	The available capacity of factory j
s_j	The production cost of factory j
T_j^{pro}	Production start time of factory j
T_i^{do}	Production end time of factory j
H_j	Warehouse volume available of factory j
E_{jr}	=1, the factory j is adjacent to the terminal r ; =0, otherwise
b_k	The unit shipping cost of LCL scheme k
T_k^{de}	The delivery time of LCL scheme k
P_{rk}^{de}	=1, the terminal of shipment for scheme k is r; =0, otherwise
T_k^{la}	Time of shipment of LCL scheme k
T_k^{sc}	Estimated time of arrival of LCL scheme k
V_k^{sc}	Available volume
D_k	Distance between terminal and warehouse of LCL scheme k
α	Cost of land transportation per unit distance
β	Production time per unit of product
h	Storage cost per unit time
σ	Penalty cost per unit time

Table 18.2 Decision variables design and description

Decision variables	Description
x_{ij}	$= 1$, small order i is produced in factory $j = 0$, otherwise
y_{ik}	$= 1$, small order i selects plan $k = 0$, otherwise

$$C^{pro} = \sum_{i-1}^{n} Q_i \times (x_{ij} \times s_j) \tag{18.2}$$

$$C^{sto} = \sum_{l=1}^{p} \sum_{i=1}^{n} V_i^{re} \times h \times (y_{ik} \times T_k^{de} - T_i^{do}) \tag{18.3}$$

$$C^{tra} = \sum_{i=1}^{n} y_{ik} \times (\alpha \times D_k + b_k \times V_i^{re}) \tag{18.4}$$

$$T_i^{Com} = y_{ik} T_k^{la} - x_{ij} T_j^{pro} \tag{18.5}$$

$$C^p = \begin{cases} \sigma \times \sum_{i-1}^{n} (T_i^{Com} - T_{\max}), \ if\ T_i^{Com} > T_{\max} & 0, \ if\ T_i^{Com} \le T_{\max} \end{cases} \tag{18.6}$$

Equation (18.1) is the objective that aims at minimizing the total cost; The C^{pro} in (2) defines the production cost, which is only related to the factory. The C^{sto} in (3) is the storage cost, which is the product of volume, unit storage cost, and storage time. The C^{tra} in (4) indicates that transportation cost, which includes land transportation and the maritime cost. Equation (18.5) is the completion time of small orders. Due to the uncertainty of shipping time, the order completion time is expressed as shipment time minus the production start time. If the actual order completion time exceeds the maximum completion time T_{max}, the penalty cost is incurred in (6).

18.3.3 Constraints

The constraints include production capacity, warehouse capacity, volume, factory restrictions, scheme restrictions, terminal, and time limits.

$$\sum_{i=1}^{m} x_{ij} Q_i \le N_j, \ \forall j \tag{18.7}$$

$$\sum_{i=1}^{m} x_{ij} V_i^{re} \le H_j, \ \forall j \tag{18.8}$$

$$\sum_{i=1}^{m} y_{ik} V_i^{re} \le \mathrm{V}_k^{sc}, \ \forall k \tag{18.9}$$

$$\sum_{j=1}^{n} x_{ij} = 1, \ \forall i \tag{18.10}$$

$$\sum_{k=1}^{K} y_{ik} = 1, \forall i \tag{18.11}$$

$$y_{ik} = \begin{cases} 0 \\ 1, \ if \ A_{ik} = 1 \end{cases}, \text{ where } A_{ik} = x_{ij} \times E_{jr} \times P_{rk}^{de} \tag{18.12}$$

$$y_{ik} \times T_k^{sc} \leq T_i^{re}, \forall i \tag{18.13}$$

Equation (18.7) ensures that the production capacity of each factory is greater than its production requirements. Equation (18.8) prevents the warehouse from overflowing. Equation (18.9) ensures that the available volume for each LCL scheme satisfies the volume of all small orders under this scheme. Equations (18.10) and (18.11) respectively indicate that only one factory and one scheme can be selected for each order. For each order selection, the restriction that the terminal of the LCL scheme near the factory is guaranteed by (18.12). If the small order takes longer than the maximum completion time, there will be a penalty cost in (18.13).

18.4 Genetic Algorithm

In this paper, the enhanced elite reservation genetic algorithm (SEGA) is used to solve the problem of production and shipping for small orders considering LCL.

The pseudo-code of the algorithm is shown (see Table 18.3).

The reciprocal method is adopted to convert the reciprocal of the objective function of feasible chromosomes into the fitness function value. At this point, the smaller value of the objective function is, the larger the fitness value is, and the solution is better. In this paper, the selected offspring is replicated with a fitness-based sorting method. Firstly, the parent population was combined with another after crossover and mutation, and the fitness of all individuals was calculated. Then, fitness values are sorted from large to small, and the first N individuals were selected for population renewal.

Chromosome coding. In this study, one chromosome represents one solution. Multi-level integer encoding is adopted, (see Fig. 18.3). A chromosome consists of the serial number of the factory and the LCL scheme selected for small orders. The chromosome length is twice the number of small orders. Small order 1 chooses factory 1 and LCL scheme 6, small order 2 chooses factory 3 and LCL scheme 5, and so on. The small order n chooses factory 2 and LCL scheme 11.

Crossover. Changing the gene sequence of chromosomes by crossing will increase population diversity. The two-point method is used here to cross. The first half of the population and the second half of the population are paired in sequence. If the number of individuals is odd, the last individual does not participate in the pairing,

Table 18.3 SEGA algorithm pseudo-code

Input:	Cost matrix, population size(pop_size), crossover probability (cro_rate), mutation probability (mut_rate) and evolution times (evo_times) of all small orders in different factory and LCL schemes
Output:	The optimal factory and LCL scheme of small orders, the optimal cost C
	initialize the population P by coding rules
	for $i = 0$ in evo_times
	calculate the fitness of the population
	perform the selection operation
	perform the crossover operation
	perform the mutation operation
	Combine the population and cross-mutated population
	perform direct copy selection based on fitness ranking
	generate new population
	end for
	The result of factories and LCL schemes is X in the population, the optimal cost C
	return X and C

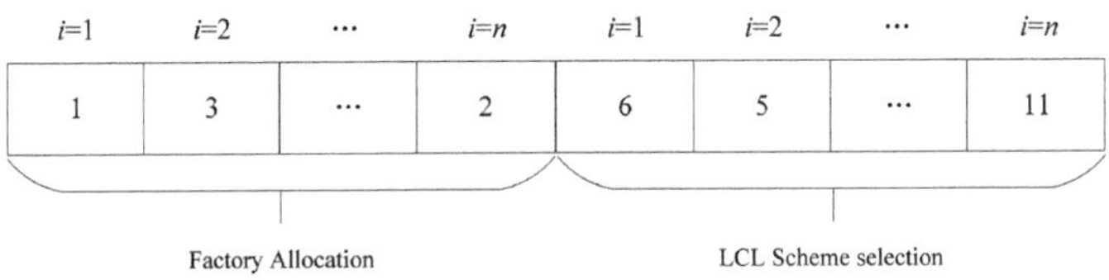

Fig. 18.3 The diagram of SEGA coding

and then the segments of each pair of chromosomes with the same genetic order are crossed (see Fig. 18.4).

Mutation. To avoid the algorithm falling into the local optimum, two random gene segments on chromosomes are exchanged with the mutation method of continuous evolution. After mutation, the characteristics of "permutation encoding" will still be here. The factory number of small order 3 has mutated. In other words, the factory number for small order 3 is regenerated (see Fig. 18.5).

18.5 Numerical Experimental Analysis

To verify the performance of SEGA in solving the problem of production and shipping for small orders considering LCL, numerical experiments are carried out in this paper. All experiments were conducted on a notebook computer equipped with AMD Ryzen

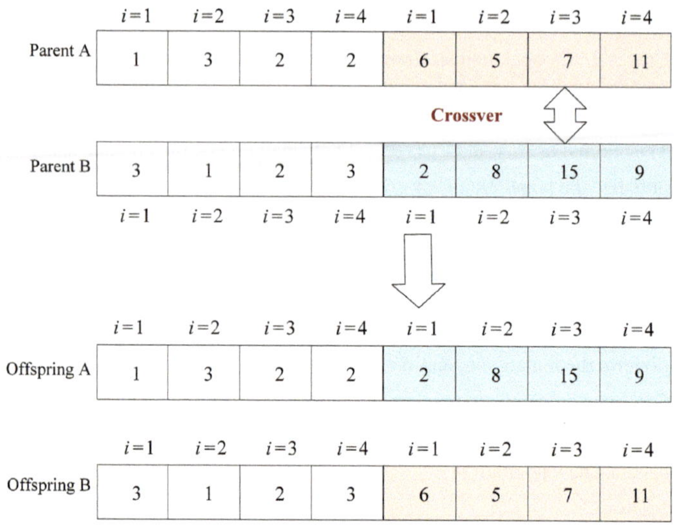

Fig. 18.4 The chromosome crossover process of SEGA

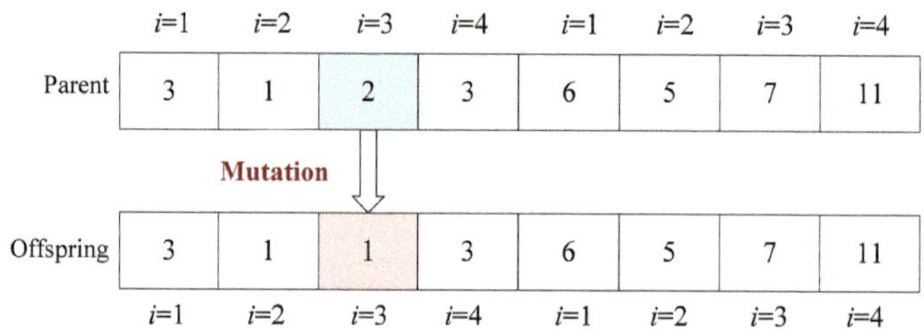

Fig. 18.5 The chromosome mutation process of SEGA

5 3550 H with Radeon Vega Mobile Gfx (2.10 GHz), and programming and numerical experiments were performed in PyCharm 2021.

It is assumed that small orders will be produced by 5 factories and shipped from 7 terminals. Meanwhile, each small order will select one from 500 LCL schemes. The parameters and their ranges are set (see Table 18.4).

The geographical locations of factories and terminals are shown (see Table 18.5). The number 1 indicates that the factory is near the terminal, otherwise the number is 0.

Table 18.4 Basic parameters and the range

	Parameters	Range of parameters
Small orders	Product quantity	5–30
	The volume required (m^3)	Product quantity × 0.35
	Required delivery time (t)	100–110
Factories	The unit cost of production	6–10
	Available production capacity	2500–3000
	Available warehouse volume (m^3)	600–650
	Production start time (t)	0–5
LCL schemes	Available volume (m^3)	40–60
	Unit shipping cost	200–300
	delivery time (t)	60–80
	Time of shipment (t)	delivery time + 2
	Estimated time of arrival (t)	80–100
	Terminal	1–7

Table 18.5 Geographical location of factories and terminals

Terminal Factory	1	2	3	4	5	6	7
1	1	1	0	0	0	0	0
2	0	0	1	1	0	0	0
3	0	0	0	0	1	0	0
4	0	0	0	0	0	1	1
5	0	0	0	0	0	0	1

18.5.1 Taguchi of Orthogonal Test

Since the solving performance of algorithms is affected by population size (pop_size), crossover rate (cro_rate), mutation rate (mut_rate), and evolution times (evo_times), the parameter values of SEGA, SGA, and studGA are optimized by Taguchi orthogonal experiment. After testing and literature review, the ranges are designed (see Table 18.6).

The table of L9 (3^3) orthogonal experiment is generated in Minitab (see Table 18.7).

Table 18.6 The range and level of SEGA algorithm parameters

Parameters	Range	Level 1	Level 2	Level 3
Pop_size	100–300	100	200	300
Cro_rate	0.5–0.9	0.5	0.7	0.9
Mut_rate	0.5–0.9	0.5	0.7	0.9
Evo_times	600–800	600	700	800

Table 18.7 The orthogonal test parameter combination of SEGA algorithm

Num	Parameters			
	pop_size	*cro_rate*	*mut_rate*	*evo_times*
1	100	0.5	0.5	600
2	100	0.7	0.7	700
3	100	0.9	0.9	800
4	200	0.5	0.7	800
5	200	0.7	0.9	600
6	200	0.9	0.5	700
7	300	0.5	0.9	700
8	300	0.7	0.5	800
9	300	0.9	0.7	600

An instance SET1 with four small orders is generated according to the rules (see Table 18.4). The SET1 was calculated 5 times under each parameter combination (see Table 18.7). The mean value can be calculated as (14). Ci in (14) is the result of the ith calculation.

$$Mean = \frac{1}{n}\sum_{i=1}^{n} C_i \tag{18.14}$$

After the experiment, the average optimal cost of SEGA at each parameter level is shown(see Fig. 18.6), which makes the best parameter levels 3, 3, 2, and 1 respectively.

Similarly, orthogonal tests are conducted on SGA and stud GA (see Table 18.8).

18.5.2 Analysis of Algorithm Performance

To verify the performance of the SEGA algorithm in solving the problem, this paper takes the SGA algorithm and stud GA algorithm as comparison algorithms and designs 5 groups of experiments to analyze the performance through the results of

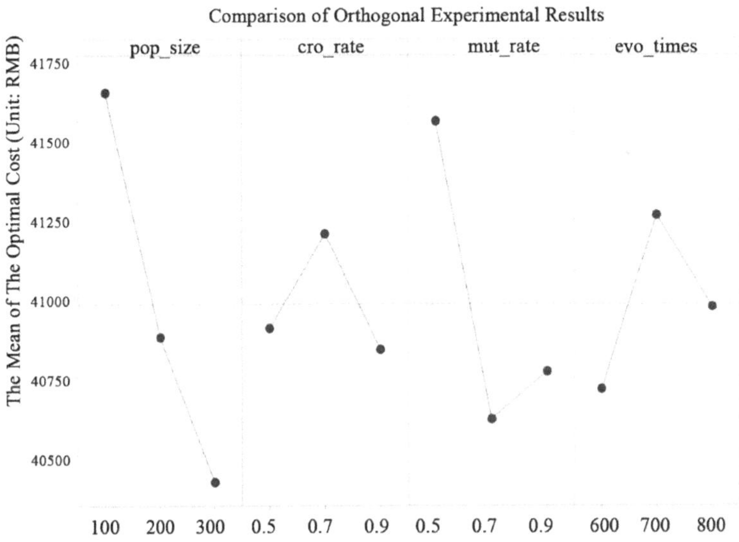

Fig. 18.6 The average optimal cost of the SEGA algorithm at each parameter level

Table 18.8 The optimal parameters of the three algorithms

	Pop_size	Cro_rate	Mut_rate	Evo_times
SEGA	300	0.9	0.7	600
SGA	300	0.5	0.9	800
Stud GA	200	0.5	0.5	600

the three algorithms at different scales. The scales of experiments, also the number of small orders, are 100, 200, 300, 400, and 500, respectively.

The data are randomly generated for each scale, and each set of data is carried out 5 times. The average results of the three algorithms under each group of experiments are counted, and the performance of the SEGA algorithm is analyzed through their differences. The experimental results are shown (see Table 18.9). The optimal cost of the SEGA algorithm is smaller than the SGA and stud GA algorithm under different scales. Compared with the average optimal cost of SGA, the values of SEGA's optimal results under the five order sizes decreased by 4.99%, 5.17%, 5.52%, 4.90%, and 3.12% respectively; compared with the average optimal cost of stud GA, the SEGA's optimal results under the five sizes decreased by 9.76%, 4.66%, 9.69%, 9.00%, and 7.57%, respectively.

Table 18.9 The average optimal cost comparison of the three algorithm under different scales (Unit: Ten thousand RMB)

Size	Num	SEGA	SGA	stud GA
n = 100	1	70.03	73.76	78.79
	2	69.75	75.08	77.67
	3	70.05	73.20	78.15
	4	70.37	74.05	77.56
	5	70.12	72.62	76.06
	Mean	**70.06**	73.74	77.65
n = 200	1	126.62	133.20	131.36
	2	125.63	133.14	130.16
	3	127.09	133.82	132.26
	4	126.32	133.22	139.84
	5	126.42	133.17	129.37
	Mean	**126.42**	133.31	132.60
n = 300	1	198.20	211.13	220.09
	2	204.32	210.16	225.61
	3	201.05	222.93	226.57
	4	203.90	211.83	223.23
	5	202.01	212.35	222.25
	Mean	**201.90**	213.68	223.55
n = 400	1	302.56	316.20	329.92
	2	301.64	315.40	334.60
	3	297.13	317.08	328.17
	4	301.50	315.92	329.85
	5	300.29	316.04	329.18
	Mean	**300.62**	316.13	330.35
n = 500	1	398.25	407.45 ara>	429.99
	2	397.22	411.07	428.92
	3	393.74	410.61	429.20
	4	395.06	408.71	429.15
	5	397.32	407.53	426.50
	Mean	**396.32**	409.07	428.75

18.6 Conclusion

As manufacturing enterprises expand overseas markets, the number of small orders is also increasing, therefore, it is extremely important to jointly optimize the production and shipping for small orders considering LCL. In this regard, this paper builds a collaborative optimization model of production and LCL shipping for overseas small-batch manufacturing needs, which is solved by the SEGA algorithm. When manufacturing enterprises allocate the factory for small orders, they select the LCL shipping scheme on the premise of meeting the time requirements. The above method provides an effective referable way for task planning of small orders.

Acknowledgements This work is supported by 'the National Key Research and Development Program of China' (2019YFE0110300).'

References

Cathy H.Y. Lam, K.L. Choy, G.T.S. Ho, and C.K.M. Lee. (2014) An order-picking operations system for managing the batching activities in a warehouse. Int J Syst Sci 45(06), 1283–1295

Feng PP, Liu Y, Wu F, Chu B (2018) Two heuristics for coordinating production planning and transportation planning. Int J Prod Res 56(21), 6872–6889

Ik Sun Lee (2015) A coordinated scheduling of production-and-delivery under dynamic delivery cost environments. Comput & Ind Eng 81, 22–35

Kuhn H, Schubert D, Holzapfel A (2021) Integrated order batching and vehicle routing operations in grocery retail—a general adaptive large neighborhood search algorithm. Eur J Oper Res 294(03), 1003–1021

Lohmer J, Lasch R (2021) Production planning and scheduling in multi-factory production networks: a systematic literature review. Int J Prod Res 59(7), 2028–2054

Li P, Li F (2021) Research on joint optimization of production and transportation considering delivery time. J Math Pract Theory 51(1), 265–276

Marandi F, Ghomi SMTF (2019) Integrated multi-factory production and distribution scheduling applying vehicle routing approach. Int J Prod Res 57(3), 722–748

Pang H (2019) Modeling and optimizing of distributed multi -factory production and transportation coordinative scheduling problem. In: Proceedings of the 2019 31st Chinese Control and Decision Conference, pp 3848–3853. IEEE, Nanchang, China

Qin J, Yang J, Dai B (2022) A joint optimization method of order allocation and path planning in e-commerce RMFS system. J Railw Sci Eng https://doi.org/10.19713/j.cnki.43-1423/u.T20220204

Sun XT, Chung SH, Chan FTS (2015) Integrated scheduling of a multi-product multi-factory manufacturing system with maritime transport limits. Transp Res Part E: Logist Transp Rev 79, 110–127

Sun XT, Chung SH, Felix TS. Chan, Wang Z (2018) The impact of liner shipping unreliability on the production–distribution scheduling of a decentralized manufacturing system. Transp Res Part E: Logist Transp Rev 114, 242–269 (2018).

Wang P, Zhang J, Yi Y (2017) Research on service efficiency optimization of electronic commerce manual parallel partition picking system. J Ind Eng Eng Manag 31(02), 209–215

Zhang B, Wang G, Yang Y, Zhang. S (2019) Solving the order planning problem at the steelmaking shops by considering logistics balance on the plant-wide process. IEEE Access 7, 139938–139956

Zhang X, Ding K. (2018) Study on order group model for multi-variety and small-batch. Chin J Syst Sci 26(01), 92–95

Chapter 19
Research on Evaluation Method of Intelligent Transportation Infrastructure Renewal Based on F-AHP Method

Lixia Bao, Jianfeng Zhou, Chuyan Zhang, Anning Ni, Yan Jiang, and Nan Ji

Abstract Due to the lack of unified evaluation standards for the renewal of transportation infrastructure, an evaluation index system for this content is constructed. Carry out intelligent evaluation on the renewal of transportation infrastructure from four aspects: smart road, smart traffic management, smart vehicle and smart network. The evaluation index is weighted by analytic hierarchy process, the threshold standard of evaluation index is determined, and the corresponding membership degree and evaluation grade are established. Then, an intelligent transportation infrastructure renewal evaluation system based on two-level fuzzy comprehensive evaluation is established. The empirical study on the renewal of transportation infrastructure in Shanghai shows that this system based on F-AHP method can reasonably evaluate the intelligent degree of transportation infrastructure renewal. The research results can guide the evaluation of urban transportation infrastructure renewal.

Keywords Intelligent transportation · Transport infrastructure renewal · Evaluation method · Analytic hierarchy process · Fuzzy comprehensive evaluation method

L. Bao · J. Zhou
Shanghai Engineering Research Center of Urban Infrastructure Renewal, Shanghai, China

C. Zhang · A. Ni (✉)
Department of Transportation Engineering, Shanghai Jiao Tong University, Shanghai, China
e-mail: nianning@sjtu.edu.cn

L. Bao · J. Zhou · Y. Jiang · N. Ji
Shanghai Urban Construction Design and Research Institute (Group) Co, Ltd, Shanghai, China

C. Yuan et al. (eds.), *Proceedings of 4th International Conference on Resources and Environmental Research—ICRER 2022*, Environmental Science and Engineering,
https://doi.org/10.1007/978-3-031-31808-5_19

19.1 Introduction

The renewal of intelligent transportation infrastructure (ITI) refers to the necessary and planned reconstruction of the infrastructure in urban transportation that can no longer adapt to the social life of modern cities into ITI with high-tech capability. It is necessary to study the evaluation indicators of ITI and how to evaluate the renewal of ITI.

This paper summarizes the research on ITI at home and abroad. In terms of ITI evaluation indicators, ISO 37151 smart community infrastructures—principles and requirements for performance metrics (Feng 2015) gives the principles and requirements for the definition, identification, optimization and coordination of urban infrastructure performance indicators, and gives analysis and suggestions on urban performance evaluation from the aspects of intelligence, interoperability, synergy, toughness and safety, but does not give specific indicators. The draft for comments on the operation index system of intelligent facilities in smart city (2022) gives the framework of the operation index system of intelligent facilities from the perspective of smart city, and explains the operation indexes of intelligent facilities. In terms of road traffic, only the operation indexes of road facilities are given. In terms of the theory and method of urban road infrastructure evaluation, Liu Jianfeng (2007) studied the theory and method of comprehensive evaluation of urban infrastructure level, and gave a solution to the specific evaluation application of "comprehensive evaluation of urban infrastructure level". Luo Qingquan et al. (2010) conducted empirical research on the index system of urban infrastructure modernization based on Chongqing, and weighted the index by entropy weight method. Huang Jinchuan et al. (2011) researched on the infrastructure construction level of 287 prefecture level and above cities in China in 2007 from the aspects of development ranking, grade division, spatial differentiation and statistical distribution by using the methods of range standardization, mean square deviation decision-making, weighted average and standard score.

The main shortcomings of the above research are as follows: (1) there is a lack of unified standard for the study of ITI renewal indicators. (2) In terms of theories and methods of urban road infrastructure evaluation, most of them study from the aspect of urban infrastructure, and there is less research on the theory of urban road infrastructure evaluation. (3) At present, the international, national, standards and local standards on smart city evaluation indicators still belong to the scope of the concept, and the infrastructure covered internally has not been refined for evaluation.

Aiming at the lack of unified evaluation criteria for the renewal of transportation infrastructure, this paper constructs an evaluation index system for the renewal of ITI using fuzzy comprehensive evaluation method, and evaluates the renewal of ITI based on F-AHP method. These methods, as suitable tools, can better solve fuzzy and difficult to the quantified evaluation problem.

19.2 Method

Fuzzy comprehensive evaluation is a method that uses fuzzy mathematics theory, combines multiple influencing factors and quantifies them at the same time, and then uses analytic hierarchy process to weight all factors. The specific evaluation process is shown in Fig. 19.1 (Zhu Boya and Zisha 2018).

19.2.1 Fuzzy Comprehensive Evaluation Method

The fuzzy comprehensive evaluation method transforms the qualitative evaluation into quantitative evaluation according to the membership theory of fuzzy mathematics. That is, the fuzzy mathematics is used to make an overall evaluation of the things or objects restricted by many factors. It has the characteristics of clear results and strong systematicness. It can better solve fuzzy and difficult to quantify problems, and is suitable for the solution of various uncertain problems (Encyclopedia and Fuzzy 2022). The basic steps can be summarized as follows:

(1) The factor domain of the evaluation object is determined.
N evaluation indexes can be set, $X = (X_1, X_2, \cdots, X_n)$.
(2) Determine the comment level domain.
Set up $A = (W_1, W_2, \cdots, W_n)$, Each level can correspond to a fuzzy subset, that is, the level set.
(3) Establish fuzzy relation matrix.

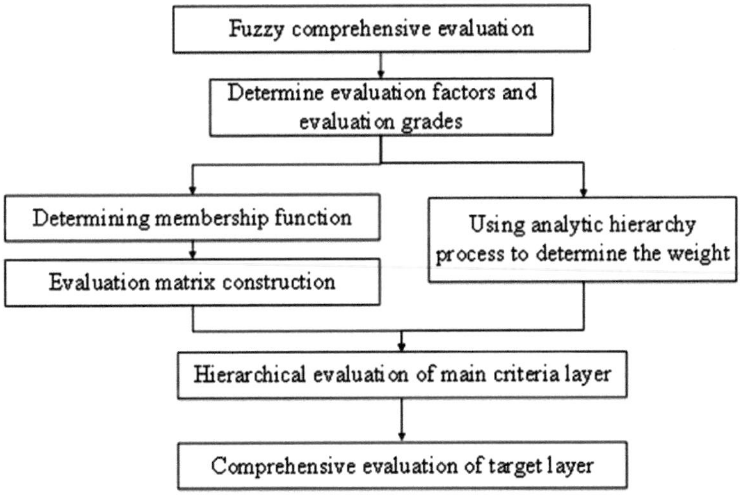

Fig. 19.1 Fuzzy comprehensive evaluation process

After constructing the hierarchical fuzzy subset, we should quantify the evaluated things one by one from each factor $X_i = (i = 1, 2, \cdots, n)$, that is, determine the membership $(\mathrm{R}|X_i)$ of the evaluated things to the hierarchical fuzzy subset from the perspective of single factor, and then obtain the fuzzy relationship matrix:

$$R = \begin{bmatrix} (\mathrm{R}|X_1) \\ (\mathrm{R}|X_2) \\ \cdots \\ (\mathrm{R}|X_n) \end{bmatrix} = \begin{bmatrix} r_{11} & r_{11} & \cdots & r_{1m} \\ r_{21} & r_{22} & \cdots & r_{2m} \\ \vdots & & \ddots & \vdots \\ r_{n1} & r_{n1} & \cdots & r_{nm} \end{bmatrix}$$

. Among them, the element in row i and column j represents the membership of a certain evaluated object X_i to the fuzzy subset of W_j grade from the perspective of factors.

(4) Determine the weight vector of evaluation factors.

In fuzzy comprehensive evaluation, the weight vector $U = (u_1, u_2, \cdots, u_n)$ of evaluation factors is determined. This paper uses analytic hierarchy process to determine the relative importance order of evaluation indexes. Thus, the weight coefficient is determined and normalized before synthesis. See Sect. 2.2 for details.

(5) Synthetic fuzzy comprehensive evaluation result vector.

The fuzzy comprehensive evaluation result vector B of each evaluated object is obtained by synthesizing u and R of each evaluated object with an appropriate operator, that is:

$$U \circ R = (u_1, u_2, \cdots, u_n) \begin{bmatrix} r_{11} & r_{11} & \cdots & r_{1m} \\ r_{21} & r_{22} & \cdots & r_{2m} \\ \vdots & & \ddots & \vdots \\ r_{n1} & r_{n1} & \cdots & r_{nm} \end{bmatrix} = (b_1, b_2, \cdots, b_m) = B$$

where b_i represents the subordinate degree of the evaluated object to the fuzzy subset of W_j level as a whole.

(6) In practice, the most commonly used method to analyze the result vector of fuzzy comprehensive evaluation is the principle of maximum membership.

19.2.2 Analytic Hierarchy Process

Analytic hierarchy process (AHP) is a qualitative and quantitative, systematic and hierarchical analysis method. The basic steps can be summarized as follows:

(1) Establish hierarchical model.

The structure diagram is generally divided into three layers, the top is the target layer, the bottom is the scheme layer, and the middle is the criterion layer or index layer.

(2) Construct pairwise comparison matrix.

Starting from the second layer, use pairwise comparison matrix and scale 1 ~ 9. If each factor in the upper layer dominates or is affected by all factors in the lower layer, it is called complete hierarchy, or called incomplete hierarchy.

Let a certain layer have n factors, $X = \{x_1, x_2, \cdots, x_n\}$. compare their influence on a certain criterion (or goal) of the upper layer, and determine the proportion of this layer relative to a certain criterion, that is, rank the influence of N factors on a certain goal of the upper layer.

The above comparison is between two factors with a scale of 1 ~ 9. To represent the comparison result of the i-th factor relative to the j-th factor, the paired comparison matrix can be constructed as $= (a_{ij})_{m\times n} = \begin{pmatrix} a_{11} & a_{11} & \cdots & a_{1n} \\ a_{21} & a_{22} & \cdots & a_{2n} \\ \vdots & & \ddots & \vdots \\ a_{n1} & a_{n1} & \cdots & a_{nn} \end{pmatrix}$.

(3) Calculate the single sorting weight vector and check the consistency.

For each pairwise comparison matrix, the maximum eigenvalue and its corresponding eigenvector are calculated, and the consistency is tested by consistency index, random consistency index and consistency ratio. If the test passes, the eigenvector (after normalization) is the weight vector; If not, the pairwise comparison matrix needs to be reconstructed.

(4) Calculate the total ranking weight vector and check the consistency.

Calculate the weight vector of the total ranking of the lowest layer to the top layer.

Use the total sorting consistency ratio $CR = \frac{a_1 CI_1 + a_2 CI_2 + \cdots + a_m CI_m}{a_1 RI_1 + a_2 RI_2 + \cdots + a_m RI_m}$, $CR < 0.1$. If yes, the decision can be made according to the result represented by the total ranking weight vector. Otherwise, it is necessary to reconsider the model or reconstruct those pairwise comparison matrices with large consistency ratio CR.

19.3 Construction of Comprehensive Evaluation Index System

The comprehensive evaluation index system is constructed from four aspects as follows.

19.3.1 Smart Road

The narrow definition mainly refers to the urban roads, including, intelligent lighting, multi pole integration and other transportation, municipal and information service facilities. In addition to the narrow connotation, the broad definition will also extend to the informatization under different road scenarios such as highway and expressway. This paper studies in a narrow sense, and the indicators include: smart lighting, smart light pole, charging station and smart bus stop.

19.3.2 Intelligent Traffic Management

Intelligent traffic management effectively applies advanced information technology, computer technology, communication technology, sensor technology, electronic control technology and artificial intelligence. The indicators include: intersection signal networking, HD video monitoring, electronic police monitoring, parking guidance system and cloud control platform.

19.3.3 Intelligent Vehicle

Intelligent vehicle integrates environmental perception, planning and decision-making, multi-level auxiliary driving and other functions. The indicators: networked vehicle terminal equipment, vehicle anti-collision system, emergency alarm system, wireless communication system, RFID vehicle management and intelligent anti-theft system.

19.3.4 Intelligent Network

Smart Internet is a integrated platform system based on the Internet of things and cloud computing, which realizes the whole process of character interaction and three-dimensional perception with the help of wireless mobile terminals, and can carry out three-dimensional composite and virtual real interaction. The indicators studied in this paper include: Road WiFi, 5G network and narrowband Internet of things.

Thus, the comprehensive evaluation index system includes 4 primary indicators and 18 secondary indicators, as shown in Table 19.1.

Table 19.1 Comprehensive evaluation system

Target layer (A)	Primary index (B)		Secondary index (C)	
	Evaluating indicator	Parameter	Evaluating indicator	Parameter
Intelligent transportation infrastructure	Smart road	B1	Intelligent lighting	C1
			Smart light pole	C2
			Charging station	C3
			Smart bus stop	C4
	Intelligent traffic management	B2	Intersection signal networking	C5
			HD video surveillance	C6
			Electronic police monitoring	C7
			Parking guidance system	C8
			Cloud control platform	C9
	Intelligent vehicle	B3	Networked vehicle terminal equipment	C10
			Vehicle anti-collision system	C11
			Emergency alarm system	C12
			Wireless communication system	C13
			RFID vehicle management	C14
			Intelligent anti-theft system	C15
	Intelligent network	B4	Road WiFi	C16
			5G network	C17
			Narrowband Internet of things	C18

19.4 Evaluation of Intelligent Transportation Infrastructure Renewal Based on F-AHP Method

19.4.1 Determination of Weight by Analytic Hierarchy Process

(1) Establish hierarchical structure model.

The structure diagram is generally divided into three layers, the top is the target layer (A), the bottom is the scheme layer (C), and the middle is the criterion layer (B). See Fig. 19.2 for details.

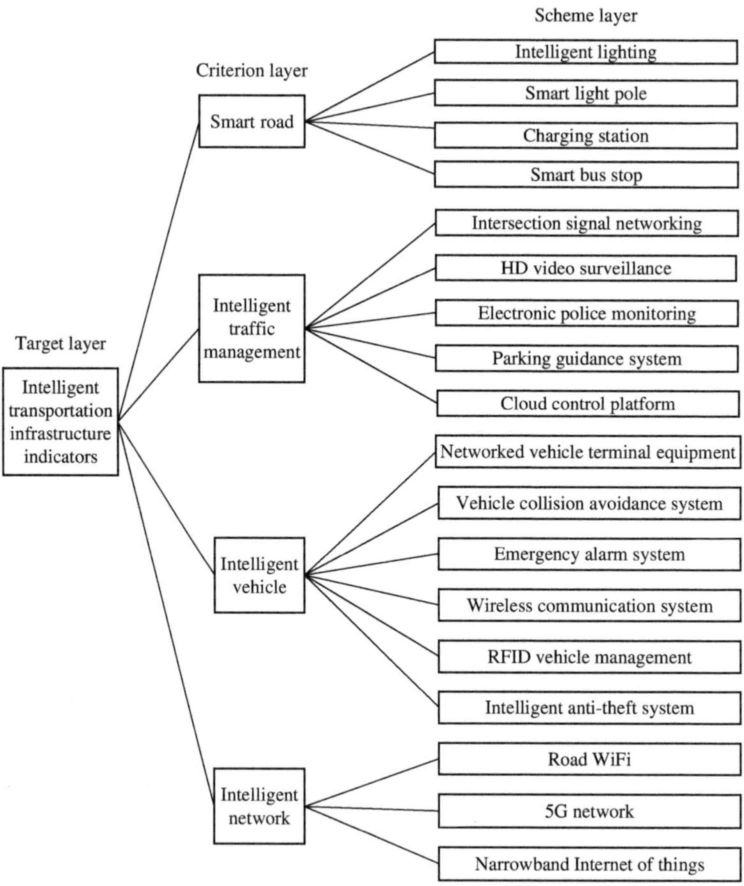

Fig. 19.2 Hierarchical analysis structure chart of ITI renewal evaluation

(2) The judgment matrix is constructed by using the scale 1 ~ 9 suggested by T.L. Saaty.

Organize experts in the field of intelligent transportation to give the relative importance of each index. The criterion layer includes smart road B1, smart traffic management B2, smart vehicle B3 and smart network B4. By comparing the relative importance of each index in pairs (see Table 19.2), a fourth-order judgment matrix A is obtained, as in (1).

$$A = \begin{bmatrix} 1 & 1/2 & 2 & 3 \\ 2 & 1 & 4 & 2 \\ 1/2 & 1/4 & 1 & 1/2 \\ 1/3 & 1/2 & 2 & 1 \end{bmatrix} \tag{19.1}$$

According to the above method, similarly, the fourth-order matrix B1, the fifth-order matrix B2, the sixth-order matrix B3 and the third-order matrix B4 can be obtained.

Calculate the sorting weight vector and check the consistency.

Eigenvalue of matrix A $\lambda_{max} = 4.1545$

One time inspection index of judgment matrix, as in (2)

$$CI = \frac{\lambda_{max} - n}{n - 1} = 0.0515 \tag{19.2}$$

$$CR = \frac{CI}{RI} = 0.057 < 0.1 \tag{19.3}$$

The value of RI is shown in Table 19.3.

Through the consistency test, the weight vector is obtained:

Table 19.2 Relative importance of primary evaluation indicators

	Smart road B1	Intelligent traffic management B2	Intelligent vehicle B3	Intelligent network B4
Smart road B1	1	1/2	2	3
Intelligent traffic management B2	2	1	4	2
Intelligent vehicle B3	1/2	1/4	1	1/2
Intelligent network B4	1/3	1/2	2	1

Table 19.3 RI values

Order	1	2	3	4	5	6	7	8	9
RI	0	0	0.58	0.90	1.12	1.24	1.32	1.41	1.45

$$w = [0.52940.76990.19250.3000]^T$$

Normalization: $w = [0.29550.42970.10740.1674]^T$

For matrices B1, B2, B3 and B4, the corresponding weights of each index are also calculated, as shown in Table 19.4 attached for detail.

Table 19.4 Weight corresponding to each index

Target layer (A)	Criterion layer (B)		Scheme layer (C)	
	evaluating indicator	Weight (w)	Evaluating indicator	Weight (h)
Intelligent transportation infrastructure A	Smart road B1	0.2955	Intelligent lighting C1	0.2284
			Smart light pole C2	0.1518
			Charging station C3	0.5189
			Smart bus stop C4	0.1009
	Intelligent traffic management B2	0.4297	Intersection signal networking C5	0.4925
			HD video surveillance C6	0.2343
			Electronic police monitoring C7	0.1588
			Parking guidance system C8	0.0707
			Cloud control platform C9	0.0438
	Intelligent vehicle B3	0.1074	Networked vehicle terminal equipment C10	0.3803
			Vehicle anti-collision system C11	0.2266
			Emergency alarm system C12	0.1510
			Wireless communication system C13	0.1161
			RFID vehicle management C14	0.0821
			Intelligent anti-theft system C15	0.0439
	Intelligent network B4	0.1674	Road WiFi C16	0.1047
			5G network C17	0.6370
			Narrowband Internet of things C18	0.2583

19.4.2 Fuzzy Comprehensive Evaluation

(1) Determine the evaluation factor set.

The smart road infrastructure renewal is divided into five categories: poor, average, medium, good and excellent, that is, the evaluation set U = {poor, average, medium, good, excellent}. The triangular function is selected to express the membership function and determine the membership degree. The membership functions determined by the corresponding evaluation set are as follows, as in (4) ~ (8).

$$u_1(x) = \begin{cases} 0 & other \\ \frac{65-x}{5} & 60 \le x \le 65 \\ 1 & x < 60 \end{cases} \tag{19.4}$$

$$u_2(x) = \begin{cases} 0 & other \\ \frac{75-x}{10} & 65 \le x \le 75 \\ \frac{x-60}{5} & 60 \le x < 65 \end{cases} \tag{19.5}$$

$$u_3(x) = \begin{cases} 0 & other \\ \frac{85-x}{10} & 75 \le x \le 85 \\ \frac{x-65}{10} & 65 \le x < 75 \end{cases} \tag{19.6}$$

$$u_4(x) = \begin{cases} 0 & other \\ \frac{90-x}{5} & 85 \le x \le 90 \\ \frac{x-75}{10} & 75 \le x < 85 \end{cases} \tag{19.7}$$

$$u_5(x) = \begin{cases} 0 & other \\ \frac{x-85}{5} & 85 \le x \le 90 \\ 1 & x > 90 \end{cases} \tag{19.8}$$

$u_1(x)$~$u_5(x)$, X represents the actual score of each evaluation index at the scheme level.

(2) First level comprehensive evaluation.

The evaluation items are divided into 4 sub items at the criterion level and 18 evaluation indexes at the scheme level. The corresponding weights are shown in Table 19.4 attached for details. Where (w_1, w_2, w_3, w_4) are the weights of the four sub items of the criterion layer, $h_1, h_2, \cdots, h_{18}$ are the weights of 18 evaluation indexes at the scheme level. According to the specific score and membership function, the first-class comprehensive evaluation J $= [J_1, J_2, \cdots, J_4]$ is obtained. R_1, R_2, R_3, R_4 represent the evaluation matrix obtained by the membership degree of the four sub item evaluation indexes of the criterion layer to the evaluation level. In this paper, the comment level is 5, the matrices are 5-column matrices, and the number of rows is the number of scheme level evaluation indicators contained in the sub items of the criterion level.

Take R_1 as an example: $R_1 = \begin{bmatrix} r_{11}\ r_{12}\ r_{13}\ r_{14}\ r_{15} \\ r_{21}\ r_{22}\ r_{23}\ r_{24}\ r_{25} \\ r_{31}\ r_{32}\ r_{33}\ r_{34}\ r_{35} \\ r_{41}\ r_{42}\ r_{43}\ r_{44}\ r_{45} \end{bmatrix}$

Where r_{ij} represents the actual score of the ith evaluation index of the scheme level and the degree of membership of the evaluation grade obtained according to the calculation of the above membership function.

The first level comprehensive fuzzy evaluation is:

$$J_i = H_i \circ R_i \tag{19.9}$$

Take J_1 as an example:

$$J_1 = H_1 \circ R_1 = (h_1 h_2 h_3 h_4) \circ \begin{bmatrix} r_{11}\ r_{12}\ r_{13}\ r_{14}\ r_{15} \\ r_{21}\ r_{22}\ r_{23}\ r_{24}\ r_{25} \\ r_{31}\ r_{32}\ r_{33}\ r_{34}\ r_{35} \\ r_{41}\ r_{42}\ r_{43}\ r_{44}\ r_{45} \end{bmatrix}$$

$$= \begin{bmatrix} j_{11}\ j_{12}\ j_{13}\ j_{14}\ j_{15} \end{bmatrix} \tag{19.10}$$

(3) Secondary comprehensive evaluation.

The first level comprehensive evaluation matrix is constructed as a second level comprehensive evaluation matrix.

$$L = W \circ J = (w_1 w_2 w_3 w_4) \circ \begin{bmatrix} J_{11}\ J_{12}\ J_{13}\ J_{14}\ J_{15} \\ J_{21}\ J_{22}\ J_{23}\ J_{24}\ J_{25} \\ J_{31}\ J_{32}\ J_{33}\ J_{34}\ J_{35} \\ J_{41}\ J_{42}\ J_{43}\ J_{44}\ J_{45} \end{bmatrix}$$

$$= (l_1, l_2, l_3, l_4, l_5) \tag{19.11}$$

(4) Fuzzy comprehensive calculation.

Normalization processing, the fuzzy comments are transformed into the total score, and the corresponding scores are given to each element in the evaluation set. In this paper, given the score distribution of each evaluation factor as {excellent = 95, good = 85, medium = 75, general = 65, poor = 55}, calculate the comprehensive score of smart road infrastructure renewal and determine the grade (Feng Chunfang et al. 2020).

$$S = L \bullet \begin{bmatrix} 55\ 65\ 75\ 85\ 95 \end{bmatrix}^T \tag{19.12}$$

19.5 Empirical Research

Taking the renewal and construction of road infrastructure in Shanghai as the evaluation object, 12 experts in the field of intelligent transportation were organized to score the evaluation indicators of renewal and construction of road infrastructure in Shanghai. Excellent: [90.100], good: [80,90), medium: [70,80), average: [60,70), poor: [50,60), and the average value is rounded to get Table 19.5 attached for details.

The scores are substituted into the membership function to obtain the membership degree of each index as follows:

$$R_1 = \begin{bmatrix} 0\ 0\ 0.1\ 0.9\ 0 \\ 0\ 0\ 0\ 0\ 1 \\ 0\ 0\ 0.9\ 0.1\ 0 \\ 0\ 0\ 0.4\ 0.6\ 0 \end{bmatrix}$$

$$R_2 = \begin{bmatrix} 0 & 0.3 & 0.7 & 0 & 0 \\ 0 & 0 & 0 & 0.8 & 0.2 \\ 0 & 0 & 0 & 0 & 1 \\ 0 & 0.7 & 0.3 & 0 & 0 \\ 0 & 0 & 0.6 & 0.4 & 0 \end{bmatrix}$$

Table 19.5 Evaluation index scores of road infrastructure renewal and construction in shanghai

Evaluating indicator		Score
Smart road B1	Intelligent lighting C1	84
	Smart light pole C2	91
	Charging station C3	76
	Smart bus stop C4	81
Intelligent traffic management B2	Intersection signal networking C5	72
	HD video surveillance C6	86
	Electronic police monitoring C7	94
	Parking guidance system C8	68
	Cloud control platform C9	79
Intelligent vehicle B3	Networked vehicle terminal equipment C10	73
	Vehicle anti-collision system C11	68
	Emergency alarm system C12	62
	Wireless communication system C13	80
	RFID vehicle management C14	86
	Intelligent anti-theft system C15	66
Intelligent network B4	Road WiFi C16	78
	5G network C17	88
	Narrowband Internet of things C18	73

$$R_3 = \begin{bmatrix} 0 & 0.2 & 0.8 & 0 & 0 \\ 0 & 0.7 & 0.3 & 0 & 0 \\ 0.6 & 0.4 & 0 & 0 & 0 \\ 0 & 0 & 0.5 & 0.5 & 0 \\ 0 & 0 & 0 & 0.8 & 0.2 \\ 0 & 0.9 & 0.1 & 0 & 0 \end{bmatrix}$$

$$R_4 = \begin{bmatrix} 0 & 0 & 0.7 & 0.3 & 0 \\ 0 & 0 & 0 & 0.4 & 0.6 \\ 0 & 0.2 & 0.8 & 0 & 0 \end{bmatrix}$$

The first level comprehensive evaluation matrix is constructed. After calculation $J_1 = \begin{bmatrix} 0 \; 0 \; 0.5302 \; 0.3180 \; 0.1518 \end{bmatrix}$. And the same calculation J_1, J_2, J_3.

Form a first-class comprehensive evaluation matrix J:

$$J = \begin{bmatrix} 0 & 0 & 0.5302 & 0.3180 & 0.1518 \\ 0 & 0.1972 & 0.3922 & 0.2050 & 0.2057 \\ 0.0906 & 0.3338 & 0.4346 & 0.1237 & 0.0164 \\ 0 & 0.0517 & 0.2799 & 0.2862 & 0.3822 \end{bmatrix}$$

Construct two-level comprehensive judgment matrix

$$W = \begin{bmatrix} 0.2955\ 0.4297\ 0.1074\ 0.1674 \end{bmatrix}$$

$$L = W \circ J = \begin{bmatrix} 0.0097\ 0.1292\ 0.4187\ 0.2433\ 0.1990 \end{bmatrix}$$

Substitute into formula (19.12) to obtain the comprehensive score and grade:

$S = L \bullet \begin{bmatrix} 55\ 65\ 75\ 85\ 95 \end{bmatrix}^T = 79.9195.$

79.9195∈ [70,80), the road infrastructure of Shanghai Smart city has been upgraded to a medium level, close to a good level.

19.6 Discussion and Conclusion

19.6.1 Discussion

1. Previous studies have not better obtained the update characteristics of intelligent transportation facilities. Through the comparison with previous studies, the established index system can better capture the update characteristics of intelligent transportation facilities.
2. The evaluation method of ITI renewal based on F-AHP method is mainly applicable to the comparison of different cities. In practical application, we can compare the intelligent degree of road infrastructure in different cities. Due to the limitation of time experience, the paper only gives the evaluation method of Shanghai, and subsequent research should evaluate the renewal of intelligent transportation facilities in different cities to illustrate the contribution of the method provided in this paper.
3. The expert evaluation used in the paper is highly subjective, and it is best to introduce big data for evaluation in the actual process.

19.6.2 Conclusion

In view of the lack of unified evaluation standards for the renewal of ITI, this paper constructs an evaluation index system for this content. Carry out intelligent evaluation on the renewal of ITI from four aspects: smart road, smart traffic management, smart vehicle and smart network. The evaluation index is weighted by analytic hierarchy process, the threshold standard of the evaluation index is determined, the corresponding membership degree and evaluation grade are established, and then the ITI renewal evaluation system based on two-level fuzzy comprehensive evaluation is established.

Through the empirical research on the renewal of transportation infrastructure in Shanghai, it shows that the renewal of road infrastructure in smart city in Shanghai belongs to the medium level and basically meets the evaluation expectations. The ITI renewal evaluation method based on F-AHP method can reasonably evaluate the intelligent degree of transportation infrastructure renewal. The research results of this paper can guide the evaluation of urban transportation infrastructure renewal.

Acknowledgements The research is supported by research project of the Science and Technology Commission of Shanghai Municipality (No.20DZ2251900).

References

Baidu Encyclopedia (2022) Fuzzy comprehensive evaluation method [EB/OL]. https://baike.baidu.com/item/%E6%A8%A1%E7%B3%8A%E7%BB%BC%E5%90%88%E8%AF%84%E4%BB%B7%E6%B3%95/2162444. 2022–05–22

Draft for Comments on Smart City (2022) Operational indicator system for intelligent infrastructure [S], 3

Feng Chunfang, Dong Kaifan, Li Zhilin (2020) Fuzzy comprehensive evaluation of expressway intelligent traffic management system based on AHP [J]. Police Technol (2):5

Huang Jinchuan, Huang Wuqiang, Zhang Yu (2011) Research on infrastructure evaluation of cities above prefecture level in China [J]. Eco Geog 31(1):8

Liu Jianfeng (2007) Research on the theory and method of comprehensive evaluation of urban infrastructure level [D] Tsinghua University

Luo Qingquan, Chen Zhongchang, Wang Qing (2010) The index system of urban infrastructure modernization is based on the empirical study of Chongqing [M]. Chongqing University Press

Yang Feng (2015) ISO 37151 smart community infrastructures—principles and requirements for performance metrics evaluation [J]. China Eco Trade Guide (8Z):4

Zhu Boya, Fu Zisha, Wang Xiaofei (2018) Traffic safety evaluation of operation area of expressway reconstruction and expansion project based on fuzzy comprehensive evaluation [J]. Highway 63(2):6

Chapter 20
Evaluation of Logistics Development Level of Wuhan City Based on Entropy Method

Huangjunshu He, Aohan Li, and Xinyu Liu

Abstract Based on the objective of building a "five type" national logistics hub city as proposed in the 14th Five-Year plan of Wuhan City, this paper constructs an evaluation index system of Wuhan City's logistics development from three aspects: economic development level, logistics scale, and logistics informatization level, which contains 18 secondary indicators, and uses the entropy method to evaluate and analyze the logistics development level of Wuhan City from 2001 to 2020. The results show that the logistics development level of Wuhan shows certain fluctuations from 2001 to 2020, but the overall growth trend is evident. Accordingly, this paper puts forward relevant suggestions in three aspects: stabilizing the development environment of Wuhan's logistics industry, strengthening the linkage between Wuhan's transportation infrastructure and logistics parks, and improving the level of logistics informatization in Wuhan.

Keywords Logistics development level · Evaluation system · Entropy method

20.1 Introduction

The 14th Five-Year Plan of Hubei Province requires Wuhan to strive to build a high-standard market system and promote the cross-regional circulation of goods and services. It also calls for improving Wuhan's energy level and efficiency in allocating factor resources in the country and accelerating the construction of a national trade and logistics center. In order to achieve these goals, it is necessary to build an evaluation index system reflecting the development level of Wuhan logistics

H. He · X. Liu
Department of Business, Wuhan Business University, Wuhan, China

A. Li (✉)
Wuhan Intelligent Logistics and Supply Chain Research Center, Wuhan Business University, Wuhan, China
e-mail: aurelialee@qq.com

C. Yuan et al. (eds.), *Proceedings of 4th International Conference on Resources and Environmental Research—ICRER 2022*, Environmental Science and Engineering,
https://doi.org/10.1007/978-3-031-31808-5_20

and to make a reasonable assessment of the development level of Wuhan logistics. Thus, the key factors affecting the development of Wuhan logistics industry can be identified, and rationalized suggestions and differentiated strategies can be put forward for a better and healthier development of Wuhan logistics industry.

Geographically, Wuhan is the central city of central China and is an essential link in China's transportation operations connecting places from east to west and north to south. For the logistics industry, Wuhan has natural geographical advantages and development prospects. The study of Wuhan's logistics development level and the promotion of the benign development of Wuhan's logistics industry will not only promote the social and economic development of Wuhan, but also provide a construction template for the development of the national logistics industry.

20.2 Establishment Evaluation of Indicators

Over the years, with the rise and growth of the logistics industry, domestic scholars have increased their research on the level of logistics development in various regions. Many scholars have conducted regionalized and characteristic evaluations of logistics development level and its influencing factors on the national logistics industry. They have constructed the evaluation index system of logistics development levels from different perspectives.

Some scholars have evaluated the level of logistics development in the country. J. Lun constructed an evaluation index system for the development level of the logistics industry for 31 provinces and cities in China (Lun 2021). G. Ma established an evaluation index system for the development of logistics in China from four aspects: transportation turnover, total communication business, increase in logistics, and average population increase (Ma 2020). X. Chen established a logistics evaluation index system for port logistics from three aspects: port condition, logistics capacity, and port economic level (Chen 2020). W. Qian, L. Wang, J. Wang and Q. Chen analyzed the spatial differentiation characteristics of logistics development levels in Chinese node cities and proposed suggestions to promote logistics development based on the measurement results (Qian et al. 2019).

Other scholars evaluate the level of logistics development for different provinces. C. Xu, H Xu, and M Qi established an evaluation system from socio-economic, infrastructure, and human resources perspectives to evaluate the development level of logistics in Shandong Province (Xu et al. 2021). M. Qiang and Q. Xiao used the fuzzy element theory to construct an evaluation index system for agricultural logistics(Qiang and Xiao 2020). Y. Wu, J. Zhu, Y. Wu, and Y. Liu established an evaluation system from four perspectives: economic development, logistics supply capacity, infrastructure, and the degree of informatization to evaluate the level of logistics development in Anhui Province (Wu et al. 2019).

In addition, some scholars have also evaluated the level of logistics development in different economic regions, such as the Yangtze River Delta region, the Beijing-Tianjin-Hebei region, and the five northwestern provinces. N. Jiang, C. Pi, and H. Jin compared the logistics development level of each port city in the Yangtze River Delta region horizontally constructed an index system, and pointed out the level differences and related suggestions in different cities from four levels: logistics operation, infrastructure, urban support, and information technology (Jiang et al. 2020). L. Sun established the index system of logistics development level in five northwestern provinces through the characteristics of logistics industry development in five northwestern provinces on "One Belt and One Road" (Sun 2020). G Wei and X Ji established an evaluation system to evaluate the level of logistics development in the Beijing-Tianjin-Hebei region from three perspectives: environment, capacity, and performance of logistics development (Wei and Ji 2019).

However, the current research mainly focuses on evaluating provinces or regions and less on the dynamic evaluation of logistics development levels for specific cities. Based on this situation, this paper establishes a rating index system for the development level of logistics in Wuhan, selecting the indicators that reflect the development of Wuhan's economy, logistics scale, and logistics informatization and establishing the evaluation index system as shown in Table 20.1.

20.3 Model Application

20.3.1 Evaluation of Logistics Development Level in Wuhan Based on the Entropy Value Method

This paper assigns weights to the indicators by the entropy value method, selects data for 20 years from 2001 to 2020, determines the importance of different indicators in logistics development, finds out the factors affecting logistics development, and improves the objectivity and comprehensiveness of the evaluation.

This paper calculates and determines the weights of each observation in the evaluation system of logistics development level based on the information provided by each observation, and the main steps are as follows.

Step 1: Set up the data matrix

$$A = \begin{pmatrix} x_{11} & \dots & x_{1m} \\ \vdots & \ddots & \vdots \\ x_{n1} & \cdots & x_{nm} \end{pmatrix} \tag{20.1}$$

$$x_{ij}{}' = \frac{x_{ij} - \min(x_{ij}, \ldots, x_{nj})}{\max(x_{ij}, \ldots, x_{nj}) - \min(x_{ij}, \ldots, x_{nj})}, \; i = 1, 2, \ldots, n,$$
$$j = 1, 2, \ldots, m, \tag{20.2}$$

Step 3: Calculate the weight of the index P_{ij}

Table 20.1 Evaluation index system of Wuhan logistics development level

First-level indicators	The secondary indicators	Unit
Level of economic development (A)	GDP (A1)	100 million yuan
	Total import and export (A2)	100 million yuan
	GDP per capita (A3)	Yuan
	Total value of imports and exports (A4)	100 million yuan
	Business volume of postal services (A5)	100 million yuan
	Total investment in fixed assets (A6)	100 million yuan
Logistics scale (B)	Total social logistics volume(B1)	100 million yuan
	Freight ton-kilometers (B2)	Ton-km
	Number of employees in transport, storage and post (B3)	10,000 persons
	Possession of civil vehicles (B4)	Thousands of cars
	Highway freight traffic (B5)	10,000 tons
	Railway freight traffic (B6)	10,000 tons
	Waterway freight traffic (B7)	10,000 tons
	Aviation freight traffic (B8)	10,000 tons
	Volume of freight handled in Ports (B9)	10,000 tons
Logistics informatization level (C)	Telecommunications revenue (C1)	100 million yuan
	Number of mobile telephone subscribers (C2)	10,000 subscribers
	Number of Internet subscribers (C3)	10,000 subscribers

$$P_{ij} = \frac{x_{ij}'}{\sum_{i=1}^{n} x_{ij}'},\ j = 1,\ 2,\ \ldots,\ m, \tag{20.3}$$

Step 4: Calculate the entropy value according to the formula

$$e_j = -\frac{1}{\ln n} \sum_{i=1}^{n} P_{ij} \ln(P_{ij}),\ \mathrm{j} = 1,\ 2,\ \ldots,\ \mathrm{m}, \tag{20.4}$$

Step 5: Calculating the coefficient of variation

$$\lambda_j = 1 - e_j,\ \mathrm{j} = 1,\ 2,\ \ldots,\ \mathrm{m}, \tag{20.5}$$

Step 6: Calculate the index weight ωj.

$$\omega_j = \frac{\lambda_j}{\sum_{j=1}^{m} \lambda_j},\ \mathrm{j} = 1,\ 2,\ \ldots,\ \mathrm{m}, \tag{20.6}$$

Step 7: Calculate the evaluation of logistics development level by linear weighted synthesis method.

$$u_i = \sum_{j=1}^{m} \omega_j \times x_{ij}{'},\ \mathrm{i} = 1,\ 2,\ \ldots,\ \mathrm{n};\mathrm{j} = 1,\ 2,\ \ldots,\ \mathrm{m}, \tag{20.7}$$

The results of the calculations are shown in Table 20.2.

Table 20.2 Comprehensive evaluation of logistics development level in Wuhan from 2001 to 2020

First level indicators	Weight	Secondary level indicators	Weight
A	0.439199158	A1	0.062620485
		A2	0.046854614
		A3	0.056520333
		A4	0.064361401
		A5	0.148810508
		A6	0.060031817
B	0.449251656	B1	0.056827005
		B2	0.044056186
		B3	0.040072208
		B4	0.071447199
		B5	0.065635701
		B6	0.033709552
		B7	0.065894195
		B8	0.043578445
		B9	0.028031165
C	0.111549186	C1	0.030667438
		C2	0.031983633
		C3	0.048898115

20.3.2 *Evaluation Results of Logistics Development Level in Wuhan*

The weight of the logistics scale and economic development level index is significant. In contrast, the weight of the logistics informatization level index is smaller, which indicates that the main driving force of logistics development in Wuhan is the economic development level and logistics scale. Therefore, we should pay attention to logistics scale input and promote economic development.

The level of logistics development and the scores of each first-level indicator during 2001–2020 were calculated based on the linear weighting method, as shown in Table 20.2. From 2001 to 2020, the total indicators of logistics development level in Wuhan are in a state of continuous growth and begin to decline in 2020.

The line graphs are plotted using the index score of each year as vertical coordinates and the year as horizontal coordinates. According to Fig. 20.1, it can be visually seen that the economic development level of Wuhan has made a qualitative leap between 2001 and 2020. Although there is a decline in 2020, it is still at a high level. The logistics scale is also steadily increasing, and although it fluctuated in 2011 and declined in 2020, the overall decline is not significant. The overall level of logistics information technology fluctuates, except for the stable growth between 2006–2015, with significant ups and downs and poor stability.

At the level of economic development, analysis of the secondary indicators reveals that the total retail sales of social consumer goods, GDP, GDP per capita, business volume of postal services, and total investment in fixed assets reached the peak in 2019 but declined with a drop of 5.36, 10.9, 23.1, 5.29, and 12.4% respectively. On the other hand, the total value of imports and exports has been growing during the 13th Five-Year Plan, with an average increase of 10.7%.

Overall, the overall economic development level of Wuhan has been hit hard by the epidemic, with various indicators declining in varying degrees compared to the previous year, which also significantly impacted Wuhan's logistics industry.

Therefore, one of the critical factors for the logistics industry in Wuhan to improve its development level after the epidemic is that we should focus on economic recovery

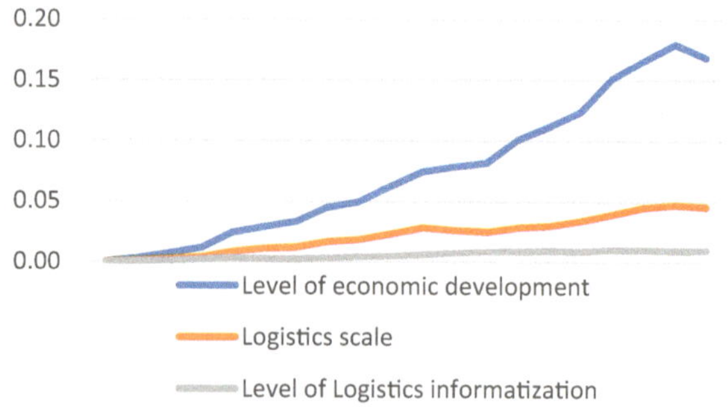

Fig. 20.1 Example of a figure caption. Comprehensive evaluation results of logistics development level in Wuhan from 2001 to 2020

and introduce policies conducive to the stable development of the logistics industry in Wuhan.

At the logistics scale level, only the total social logistics and highway freight traffic data change relatively steadily during 2001–2020 for each secondary indicator, with slight overall fluctuation, while the other indicators fluctuate to different degrees, showing the instability of their development. Overall, the scale of logistics in Wuhan is developing well. Even during the 2020 epidemic, most indicators such as freight ton-kilometers, possession of civil vehicles, waterway freight traffic, volume of freight handled in Ports can still maintain growth. The indicators of the freight traffic accomplished by air, rail, and road and the number of employees in the transport, storage, and postal industry all declined.

The decline is not only due to the blow of the epidemic but also to the inadequate development of Wuhan's logistics industry.

The logistics layout of Wuhan is still under construction, which makes the linkage between various modes of transportation and the surrounding logistics parks insufficient and thus less able to cope with unexpected situations, which is one of the critical factors limiting the development of Wuhan's logistics industry.

The level of logistics informatization has significant volatility in the score between 2001 and 2020, indicating its poor stability. The overall score peaked in 2017 and then declined for two consecutive years, with an average decrease of 4.24%. The indicator still managed to buck the trend in 2020 with an increase of 2.17% under the impact of the COVID-19 epidemic.

The main reason for its growth is that during the epidemic, various activities were switched to online, which increased the use of network communication on the one hand and improved the overall level of logistics informatization on the other. It shows that the level of logistics informatization in Wuhan is still not high, the development stability is low and needs to be strengthened, and the improvement effect is immediate.

20.4 Conclusions and Suggestions

20.4.1 Conclusions

The level of logistics development in Wuhan fluctuated to a certain extent in 2020, but the overall growth trend is obvious. Among them, the level of logistics informatization fluctuates wildly, there is no obvious upward or downward trend, and the stability is not strong, which needs to be paid great attention to.

The current development of the logistics industry in Wuhan still suffers from an unstable development environment, insufficient linkage between logistics park and infrastructure, and a low level of logistics informatization. This paper provides reference suggestions regarding policy, logistics infrastructure, and logistics informatization construction.

20.4.2 *Suggestion*

For the overall development of Wuhan's logistics industry, this paper puts forward the following recommendations:

Policies to help the development of the logistics industry in the Post-epidemic era, stabilizing the general environment for the logistics development. Through government support, policy assistance, and cooperation with universities, the city's own geographical and talented advantages should be used to attract large logistics enterprises to build logistics parks and other facilities in Wuhan. On the one hand, we can improve the construction level of Wuhan's logistics industry infrastructure. On the other hand, we can highlight the advantages of Wuhan's logistics development so that it is no longer vulnerable to the influence of the incident.

Accelerate the construction of logistics parks and strengthen the linkage between transportation infrastructure and logistics parks. The government can use logistics parks such as East–West Lake Comprehensive Logistics Park and Yangluo Port, combined with Wuhan's expressways, railroad trunk lines, and airports, to form a four-level spatial layout of the logistics industry and enhance the ability of Wuhan's logistics industry to radiate to the national as well as international markets.

Strengthen the logistics informatization construction level and improve stability of logistics information development. The informatization and standardization construction of logistics parks should be Accelerated.

For example, the government joint logistics information service platform to promote the interconnection of information between logistics platforms to ensure the stability of the development of logistics information level; the government can also promote the construction of the standard system to improve the overall logistics information level.

Acknowledgements The research works in this dissertation are financially supported by the project "Research on the evaluation of high-quality logistics development of urban agglomeration in the middle reaches of the Yangtze River " (Project Number: SKKT-22067).

References

Chen X (2020) Evaluation of China's development level of port logistics indicators and countermeasures research. Logis Eng Manag 42(3):7–8

Jiang N, Pi C, Jin H (2020) Multidimensional evaluation of logistics development level in Yangtze River Delta port cities. China Stor Transp 6:114–115

Lun J (2021) Research on a comprehensive evaluation of logistics industry development level of 31 provinces and cities in China. Invest Entrep 32(3):50–52+55

Ma G (2020) Evaluation of development level of logistics industry in China and analysis of spillover effect. Busi Econ Res 11:126–129

Qian W, Wang L, Wang J, Chen Q (2019) Research on the logistics development level and spatial differentiation in node city with gray target decision model. Grey Syst Theory Appl 11(1):183–199

Qiang M, Xiao Q (2020) Evaluation of Gansu rural logistics development level based on fuzzy matter-element theory. Log Techn 39(12):26–32

Sun L (2020) Evaluation of the development level of the logistics industry in five provinces (regions) of Northwest China in the context of "One Belt and One Road." J Qinghai Normal Uni (philosophy and Social Science Edition) 42(1):21–27

Wei G, Ji X (2019) Comprehensive evaluation of logistics development level in Beijing-Tianjin-Hebei region. Busi Eco Res 16:101–103

Wu Y, Zhu J, Wu Y, Liu Y (2019) Comprehensive evaluation of anhui logistics development level based on factor analysis and cluster analysis. J Wuhan Light Ind Uni 38(6):74–80

Xu C, Xu H, Qi M (2021) Research on a comprehensive evaluation of development level of urban logistics in Shandong province. J Nanyang Ins Techn 13(4):1–9